ASE Study Guide **by Chek-Chart**

Brakes

D1792066

FOR ASE TEST A5

Contents

Brakes, ASE Study Guide **by Chek-Chart**
Copyright © 1998 by Chek-Chart Publications

All rights reserved. Printed in the United States of America. No part of this publication may be reproduced, stored in a retrieval system, or duplicated in any manner without prior written permission of the publisher.

International Standard Book Number: 1-57932-096-1

Printed in the United States of America.

Chek-Chart Publications
320 Soquel Way
Sunnyvale, CA 94086
408-739-2435
"Chek" us out on the web at www.chekchart.com.

Editor
William J. Turney, CMAT, L1

Contributing Editor
Richard K. DuPuy

Technical Consultant
Jerry Mullen

Production Team Supervisor
Tricia Flodder

Production Team
Svetlana Dominguez
Maribeth Echard
Kristy Nash
Carl Pierce
Mary Ellen Stephenson

DISCLAIMER

Chapter One

HYDRAULIC SYSTEM SERVICE

Hydraulic brake system service includes diagnosing and correcting problems that relate to brake fluid condition and level, master cylinder operation, brake fluid lines and hoses, and the hydraulic control valves that regulate system pressure. These services, as well as bleeding air from the hydraulic system, are discussed in this chapter.

BRAKE FLUID

When topping off the brake fluid, always use the type and DOT grade of brake fluid recommended by the manufacturer. Most vehicles use polyglycol brake fluid, either a DOT 3 or DOT 4 grade fluid. Ford vehicles require a special DOT 3 with an extremely high boiling point. Many import vehicles use DOT 4. Occasionally, you will run across a vehicle that uses DOT 5 or silicone brake fluid. Some European manufacturers, such as Citroen and Rolls-Royce, use Hydraulic System Mineral Oil (HSMO) in the brake system, while Audi uses it in the hydraulic brake booster on select models.

To avoid intermixing the three types, federal law requires that each be a specific color: polyglycol fluids are clear to amber, silicone fluids are purple, and hydraulic mineral oils are green.

Storage and Handling

Storage and handling precautions for brake fluid depend on the type of fluid being used. Polyglycol fluid has a very limited storage life. Once a can of polyglycol fluid has been opened, its entire contents should be used as soon as possible because it immediately begins to absorb moisture that degrades its performance. In contrast, silicone brake fluid and HSMO can be stored almost indefinitely. They are not **hygroscopic**, and there is no limit to the length of time they retain their original properties.

When handling a polyglycol fluid, remember that it is a powerful solvent that can rapidly damage paint. If you spill any, immediately flush the area with plenty of

clean water. Neither silicone nor HSMO brake fluids harm paint.

Brake Fluid Condition

Whenever you check the brake fluid level, also inspect the fluid for dirt, moisture, or oil contamination. Fluid in good condition should be relatively clear. A cloudy appearance indicates moisture contamination, while a dark appearance indicates contamination by rust, dirt, corrosion, or brake dust. A layered appearance can mean a silicone brake fluid was mixed with a polyglycol fluid. The two fluids do not mix, and the entire system should be completely flushed, then refilled with the recommended fluid.

MASTER CYLINDER SERVICE

Master cylinder service begins with a visual inspection:

1. Ensure fluid is at the proper level.
2. Make sure the reservoir cover vent holes are clean and unrestricted.
3. Look over the master cylinder diaphragm, if there is one, for any cracks, tears, or other damage.
4. Check for external leaks at line connections or at the pushrod.
5. Inspect dust boots, if used; they must be soft and without cracks, and there should be no fluid behind them.

Master Cylinder Testing

A number of special tests can pinpoint whether a problem is with the master cylinder or elsewhere in the system. First, apply the brake pedal to make sure there is the correct amount of freeplay, figure 1-1. Most systems require between 0.13 and 0.50 inch (3 and 13 mm) of freeplay.

Check pedal feel. A spongy pedal with longer than normal travel usually indicates a hydraulic problem, such as a fluid leak or air in the lines. If the brake pedal gradually sinks part way to the floor, then becomes

---

Hygroscopic: Water absorbing. Polyglycol brake fluids are hygroscopic.

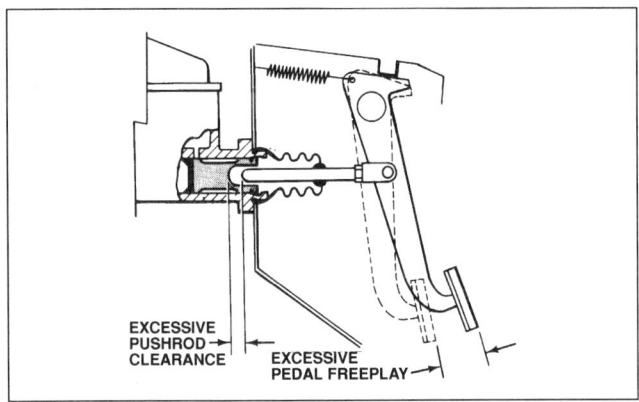

1-1 *The first step in master cylinder testing is to check for correct brake pedal freeplay.*

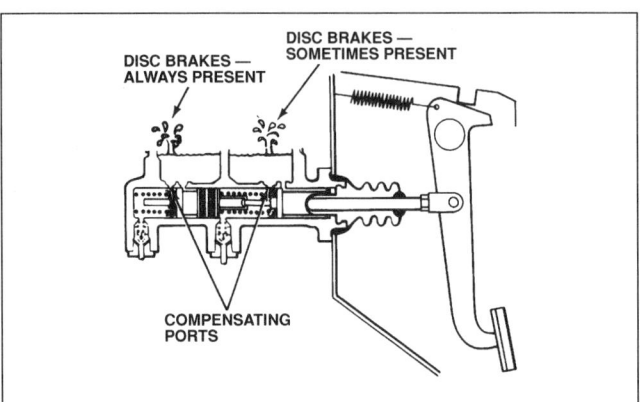

1-2 *A properly functioning, or open, compensating port allows a small spurt of fluid in the reservoir as the brake pedal is applied.*

firm, one circuit of the dual-circuit brake hydraulic system is probably at fault. If the pedal gradually sinks all the way to the floor, suspect a master cylinder that is bypassing internally.

External leak test

A low master cylinder fluid level indicates either normal brake lining wear or a hydraulic system leak. Make sure the cylinder is at least half full, and note the exact level. If the cylinder has run dry, bleed the system first. Apply the brake several times, then check the level again. If the fluid level dropped, there is an external leak.

Internal leak test

An internal leak test, also called a bypass test, checks the integrity of the primary seals on the master cylinder piston. The cylinder reservoir must be at least half full for testing. To test, watch the fluid in the reservoir as an assistant slowly applies and releases the brake pedal. If the fluid level rises as the pedal is applied and falls as it is released, the seals are leaking. The fluid is bypassing the seals and rising into the reservoir.

Compensating port test

If there is no pedal freeplay, the **compensating ports** in the master cylinder may be closed. To test, remove the reservoir cover and watch the fluid in the reservoir as an assistant slowly applies the brakes. A small amount of fluid should be forced out of the cylinder bore through the compensating port in each chamber. This causes a small spurt of fluid on the surface, figure 1-2.

When performing the compensating port test on a **quick-take-up (QTU) master cylinder**, the **quick-take-up (QTU) valve** initially restricts fluid flow through the rear compensating port, so a jet of fluid will not appear in the reservoir. However, once the clearance in the brake system is taken up and pressure reaches about 70 to 100 psi (483 to 690 kPa), the QTU valve check ball unseats, and a large quantity of fluid is pumped into the reservoir very rapidly. This can create a safety hazard, so have your assistant apply the brake pedal very lightly. Hand pressure is best, so the opening pressure of the QTU valve is not exceeded.

Quick-take-up valve test

This valve regulates the flow of fluid between the reservoir and the master cylinder chambers. While it cannot be directly tested, a problem may be indicated if there is excessive pedal travel when the brakes are first applied or if the brake pedal returns slowly when released.

Master Cylinder Replacement

To remove a master cylinder:

1. Use a flare nut wrench to disconnect the brake lines from the master cylinder fluid outlets. Plug the outlets and cap off the open lines to prevent fluid spillage and system contamination.
2. If there is a fluid level warning switch on the reservoir or a pressure differential switch on the master cylinder body, detach the wiring harness connector.

Compensating Port: The opening between the master cylinder reservoir and the cylinder bore that allows fluid to enter or exit the hydraulic system to adjust for changes in volume.

Quick-take-up (QTU) Master Cylinder: A type of master cylinder that applies a large volume of fluid on the initial brake application to take up the clearance designed into low-drag brake calipers.

Quick-take-up (QTU) Valve: The part of the QTU master cylinder that controls the fluid flow between the reservoir and the cylinder bore.

3. If the vehicle has manual brakes and the brake pedal pushrod is mechanically connected to the master cylinder, disconnect the pushrod from the brake pedal linkage.
4. Remove the bolts that attach the master cylinder to the firewall or power booster, then lift the cylinder from the vehicle. Keep the cylinder upright to prevent fluid spillage.

Once the cylinder is removed from the vehicle, make an internal inspection to determine whether the cylinder can be rebuilt. Internal inspection and rebuild procedures are covered later in this chapter. Install the rebuilt or new replacement master cylinder as follows:

1. Bench bleed the master cylinder and cap off the fluid ports.
2. Fit the master cylinder on the firewall or power booster, install the mounting bolts, and tighten them to specified torque.
3. Connect the brake lines to the master cylinder, start the threads by hand, then tighten the fittings with a flare-nut wrench.
4. Attach the wiring harness connector to the fluid level warning switch or pressure differential switch if the cylinder is so equipped.
5. Connect the brake pedal pushrod to the pedal linkage if it was detached when the cylinder was removed.
6. Check and adjust the pedal freeplay and mechanical stoplight switch, then bleed the system. These procedures are detailed later in this chapter.

Pedal Linkage Adjustment

It is essential to adjust the pushrod on a new or reconditioned master cylinder to establish the correct brake pedal free play. To adjust the pushrod:

1. If the vehicle has a power booster, pump the brake pedal until the reserve is exhausted and the pedal feel hardens.
2. Place a ruler along the axis of brake pedal travel, then slowly apply the pedal by hand until all the slack in the linkage is eliminated. This amount of travel is the freeplay.

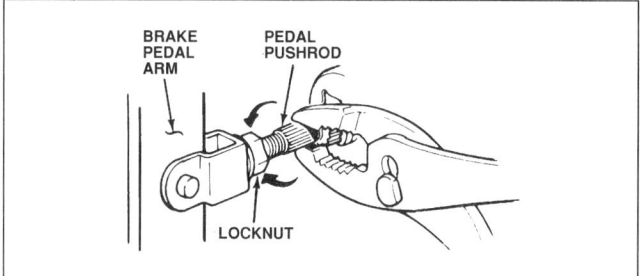

1-3 Brake pedal freeplay is usually adjusted by loosening a locknut and turning the pushrod to obtain the proper length.

3. Adjust the freeplay by shortening or lengthening the brake pedal pushrod. Loosen the locknut on the pushrod, figure 1-3. Rotate the pushrod until you get the specified free play, then tighten the locknut.

Master Cylinder Overhaul

Overhauling the many and various master cylinders differs, so it is best to consult the vehicle shop manual when overhauling a master cylinder. The following procedures are common practices and general rules that apply to overhauling most master cylinders.

Disassembly

To disassemble a master cylinder:

1. Remove the pushrod and reservoir cap, then drain any remaining fluid from the fluid reservoir.
2. If there is a dust boot on the pushrod end of the cylinder, remove it.
3. Clamp the master cylinder in a vise by its mounting flange. If the fluid reservoir is not part of the cylinder body, remove it.
4. If the cylinder body has tube seat inserts, thread self-tapping screws into the seats, then pry them out using two screwdrivers. Drum brake master cylinders may have residual pressure check valves beneath the seats, which you can now remove.
5. If the pressure differential switch or proportioning valves are threaded into the master cylinder body, remove them. Do not remove the QTU valve unless testing indicates that it needs replacing.
6. Slightly depress the primary piston, then remove the snapring which acts as a piston stop. Some master cylinders have a Allen-head plug that must be removed first.
7. Withdraw the stop washer, primary piston, spring, primary cup, and valve seat from the master cylinder, figure 1-4. If the master cylinder has a stop bolt for the secondary piston, remove it. Use low-pressure compressed air to remove stubborn pistons.
8. Clean the entire cylinder with brake parts cleaner, alcohol, or fresh brake fluid.
9. Inspect the housing casting for cracks, damaged threads, or other signs of damage.
10. Inspect the compensating intake ports; they must be clean and open.

Internal inspection

Shine a light into the cylinder and inspect the bore. If the unit is made of cast iron and the bore is in good condition or only lightly scratched, pitted, scored, or

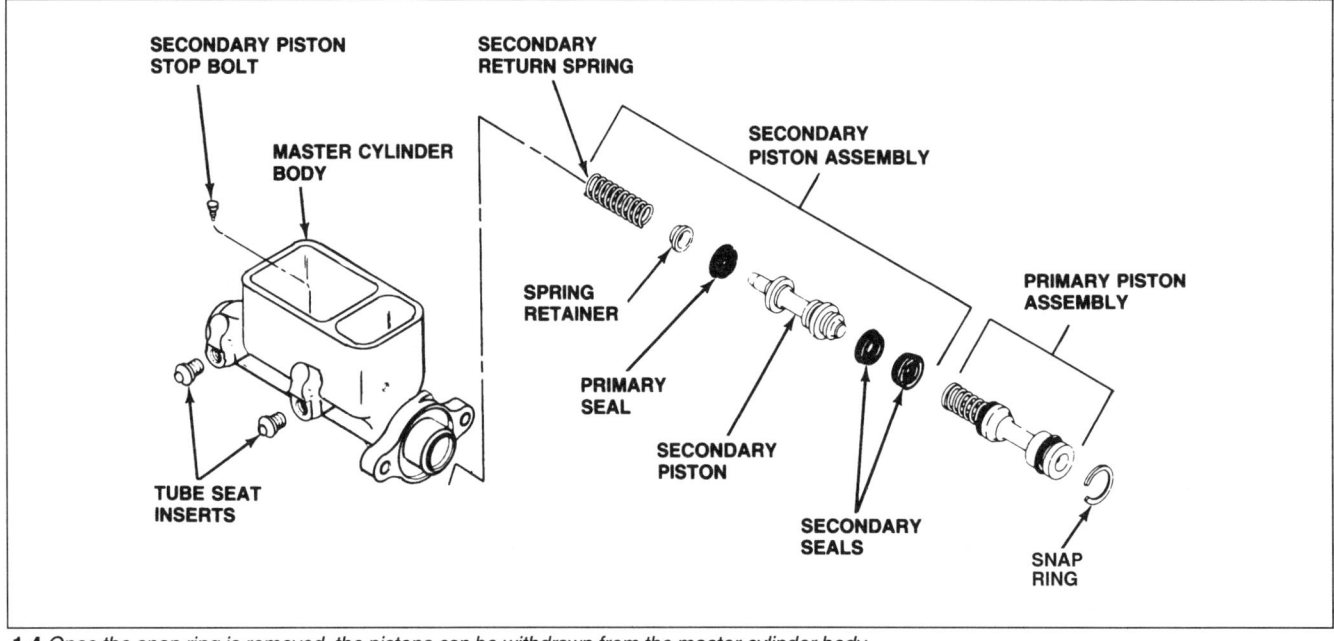

1-4 *Once the snap ring is removed, the pistons can be withdrawn from the master cylinder body.*

corroded, the cylinder can probably be honed and re-built. Be aware, some manufacturers recommend against honing, and it is advantageous to check the shop manual for specific recommendations.

If the bore is deeply scored, the cylinder must be replaced. Honing to remove deep scores results in an oversized cylinder bore, and the new seals cannot withhold pressure, which causes internal leakage.

If an aluminum hydraulic cylinder is scratched, pitted, scored, or corroded in any way, the cylinder must be replaced. Aluminum cylinders cannot be honed because they have a wear-resistant, **anodized finish**, which honing removes.

Honing

Use the following procedure to hone the mildly damaged bore of a cast-iron master cylinder:

1. Clamp the cylinder in a vise by its mounting flange.
2. Select a suitably sized cylinder hone and chuck it in a drill motor.
3. Lubricate the cylinder bore with fresh brake fluid and insert the hone into the bore.
4. Operate the drill motor at approximately 500 rpm, and move the hone back and forth along the entire length of the bore using smooth, even strokes. Never allow the hone stones to come partially out of the end of the bore while honing.

5. Keep the bore lubricated with fresh brake fluid and hone for approximately 10 seconds. Allow the hone to come to a full stop before removing it from the bore.
6. Rinse the bore with fresh brake fluid, wipe it clean with a rag, and check the surface finish. It should be free of rust, corrosion, and scratches. The hone should also have created an even crosshatch pattern, figure 1-5.
7. If necessary, repeat the honing process for another 10 seconds. If the bore does not clean up after several repetitions of this procedure, replace the cylinder.

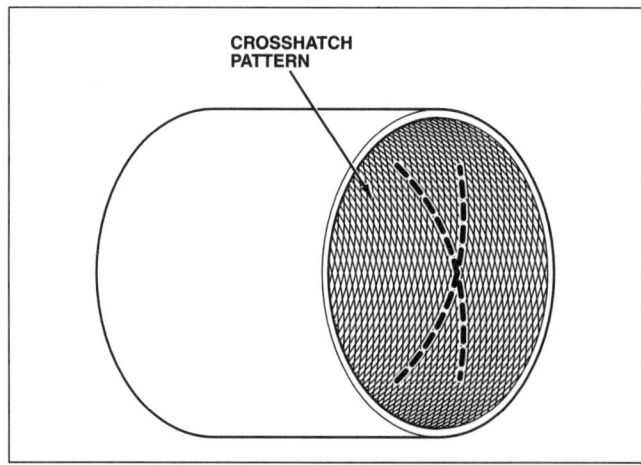

1-5 *A properly honed cylinder will have an even crosshatch pattern that promotes seal seating.*

Anodized Finish: An electrolytically applied coating of a protective oxide. It slows wear.

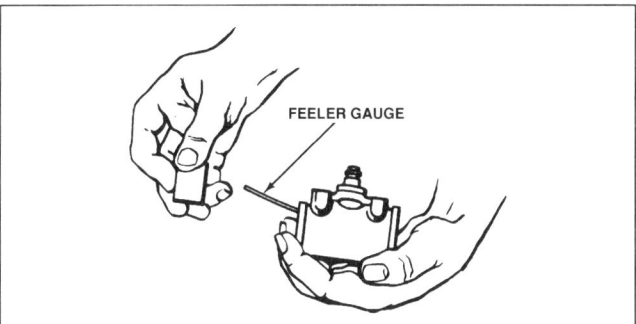

1-6 *If the piston can be inserted into the bore along with a 0.006-inch (0.15 mm) feeler gauge blade, the bore is oversized, and the cylinder must be replaced.*

8. After the bore is honed, thoroughly clean the cylinder with a non-petroleum–base brake cleaning solvent to remove all residue and grit.

Cylinder bore measurement

After honing a hydraulic cylinder, measure the bore diameter to make sure that too much metal has not been removed. There are two methods of measuring a cylinder bore: One uses a feeler gauge strip, and the other uses a go/no-go gauge.

To check the bore using a feeler gauge, place a narrow, 0.02 inch (6 mm) wide, strip of 0.006-inch (0.15 mm) feeler gauge inside the bore. Attempt to insert one of the cylinder pistons into the bore with the feeler gauge in place. If the piston fits, the bore is oversized, and the cylinder must be replaced, figure 1-6. Traditionally, most manufacturers have allowed up to 0.006 inch (0.15 mm) of piston clearance. However, many newer cylinders with smaller diameters require tighter clearances. If you are unsure of the proper specification, consult the factory shop manual.

To check the bore with a go/no-go gauge, select the proper sized plug from the gauge kit and attach it to the handle. The correct plug is the same size as the nominal cylinder bore plus 0.006 inch (0.15 mm).

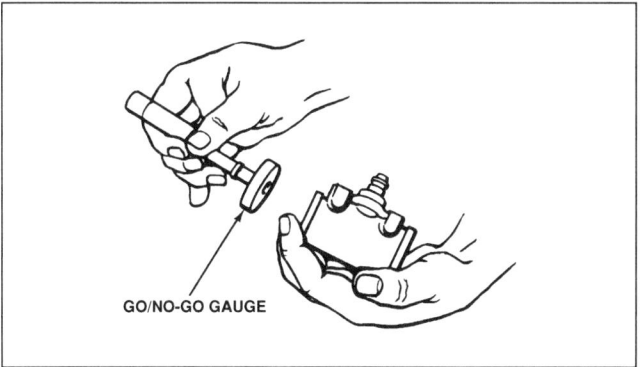

1-7 *A correct go/no-go gauge is sized to the maximum diameter of the cylinder bore. If the gauge fits into the bore, the bore is oversized and the cylinder must be replaced.*

Attempt to insert the plug into the bore. If the gauge fits into the cylinder, the bore is oversized, and the cylinder must be replaced, figure 1-7.

Reassembly

To reassemble a master cylinder:

1. Install new residual check valves into the fluid outlet ports, if the cylinder was so equipped. Drive new tubing seats in over the check valves using a brass drift or wooden dowel.
2. Assemble the return spring, spring retainer, and seals onto the secondary piston. Be sure to install seals in the right direction; the seal lip must face toward the hydraulic apply pressure.
3. Lubricate the secondary piston assembly with clean brake fluid and fit it, spring end first, into the bore. If a piston stop bolt is used to hold the piston in place, install it.
4. Assemble the return spring, spring retainer, and seals onto the primary piston. Lubricate the primary piston assembly with fresh brake fluid and install into the bore spring end first.
5. Depress the primary piston and fit the retaining snapring. Install the Allen-head plug, if equipped.
6. Install the fluid reservoir, pressure differential switch, and proportioning valves as required.
7. Bench bleed the master cylinder assembly as described later in this chapter. Then, install the rebuilt cylinder on the vehicle.

BRAKE BLEEDING

Brake bleeding is a process that pushes new brake fluid through the brake system to force out contaminated fluid and trapped air. Air can enter the brake lines whenever the system is opened for service. Brake system bleeding must be done in a particular sequence:

1. Master cylinder
2. Combination valve
3. Wheel cylinders and brake calipers
4. Load-sensing proportioning valve
5. Antilock brake system (ABS) hydraulic modulator or pump motor.

Combination valves, load-sensing proportioning valves, and ABS hydraulic modulators or pump motors are only bled when present in the system and equipped with bleeder valves. The correct bleeding sequence at the wheels varies from vehicle to vehicle. In some cases, you may have to recenter the pressure differential switch after bleeding the wheel brakes. The centering procedure is detailed later in this chapter.

Master Cylinder Bleeding

Bleed new or rebuilt cylinders on the workbench before installing them. This practice eliminates trapped air pockets and greatly speeds bleeding the rest of the brake system.

Master cylinder bench bleeding

1. Clamp the cylinder in a vise.
2. Use a drain pan to catch any fluid leakage from the outlet ports, or fit hoses to the ports to route fluid back to the reservoir. Fill the cylinder with the correct type and grade of new brake fluid.
3. Use the cylinder pushrod, or other round-ended rod, to slowly stroke the cylinder pistons inward until they both bottom.
4. If the fluid outlets are open, plug them with your fingertips. Then, slowly allow both pistons to fully return on the back stroke and remove your fingers from the outlets. If the fluid outlets are connected to the reservoir with hoses, make sure the connections are airtight. Then, slowly allow both pistons to fully return.
5. Repeat steps 3 and 4 until the fluid coming from the outlets is air-free, and bubbles no longer emerge from the compensating and replenishing ports in the reservoir.

Wheel Brake Bleeding

When bleeding the wheel brakes, follow the bleeding sequence recommended by the manufacturer for the particular vehicle being serviced. Sequence can vary not only by manufacturer, but also by year, model, and equipment. Special procedures may be required for vehicles with ABS. Generally, the wheel cylinder or caliper furthest from the master cylinder is bled first, followed by the next closest caliper or cylinder, and so on. Be sure to check the recommendations for the specific vehicle you are working on, as sequences will vary. In all types of wheel brake bleeding, fill the master cylinder with fluid and ensure that it stays at least half full during the entire procedure.

Manual brake bleeding

This method requires two people: One person to press the brake pedal, while the other opens and closes the bleeder valves. The brake pedal will sink toward the floor as the valve is opened. Upon signal, the bleeder valve is tightened and the operation repeated until all the air is expelled. To manually bleed a system:

1. If the vehicle has a vacuum or hydraulic power booster, discharge it by pumping the brake pedal with the ignition OFF until the pedal feel hardens.
2. Slip a length of clear plastic hose over the bleeder valve of the first wheel cylinder or caliper in

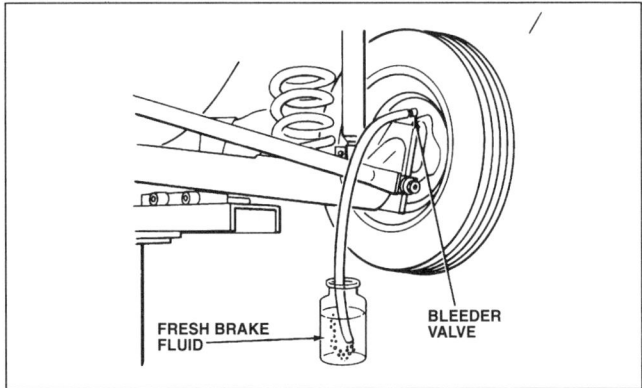

1-8 *Using clear hose and a partially filled bottle of fresh brake fluid makes it easy to spot air bubbles and prevents spills when bleeding the system.*

the bleeding sequence, and submerge the open end of the tube in a partially filled container of fresh brake fluid, figure 1-8.
3. Loosen the bleeder valve approximately one-half turn, then have your assistant slowly depress the brake pedal and hold it to the floor. Air bubbles leaving the bleeder valve will be visible in the hose to the container.
4. Tighten the bleeder valve, then have your assistant slowly release the brake pedal.
5. Repeat steps 3 and 4 until no more air bubbles emerge from the bleeder valve.
6. Transfer the plastic hose to the bleeder valve of the next wheel cylinder or caliper in the bleeding sequence, and repeat steps 3 and 4. Continue around the vehicle in the specified order until the brakes at all four wheels have been bled.

Pressure tank bleeding

Pressure bleeding drum brake systems is a straightforward procedure, but some disc brake systems have a metering valve or combination valve that demands special attention. Since the operating pressure of power bleeders is within the range where the metering valve blocks fluid flow to the front brakes, you must deactivate the valve. Special override tools are used to hold the valve open for pressure bleeding, figure 1-9.

Follow the correct sequence when pressure bleeding a brake system. Be aware, some manufacturers recommend one sequence for manual and another for pressure bleeding. To pressure bleed a system:

1. Make sure the pressure bleeder tank is filled with the proper type and grade of brake fluid. Consult the instructions for the equipment being used.
2. With the bleeder properly sealed and the fluid supply valve closed, use compressed air to charge the bleeder to approximately 30 psi (207 kPa) of pressure.

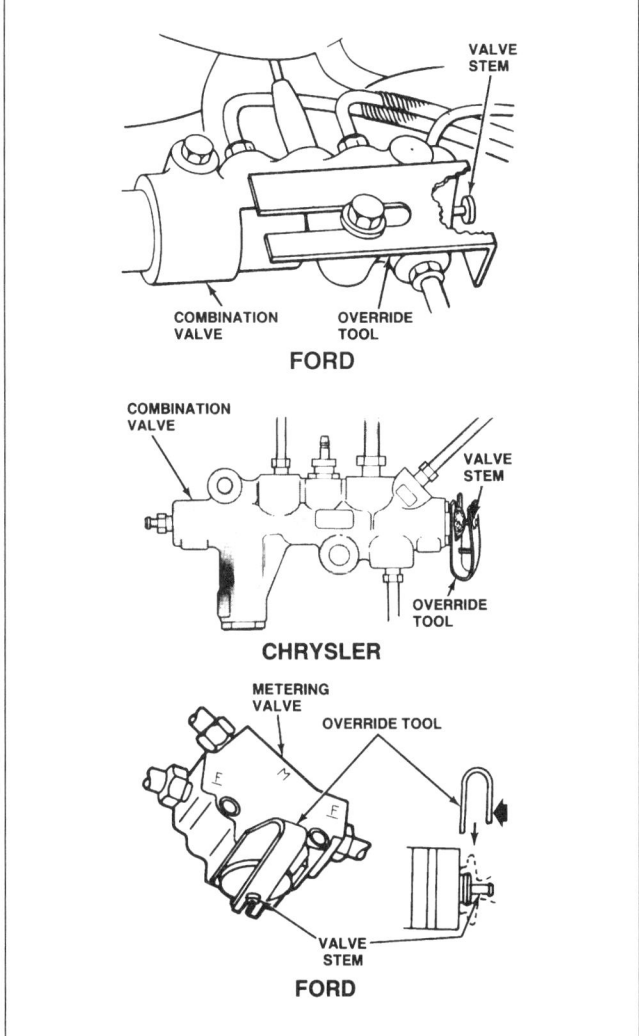

1-9 *Override tools hold the combination or metering valve open to allow for pressure bleeding the hydraulic system.*

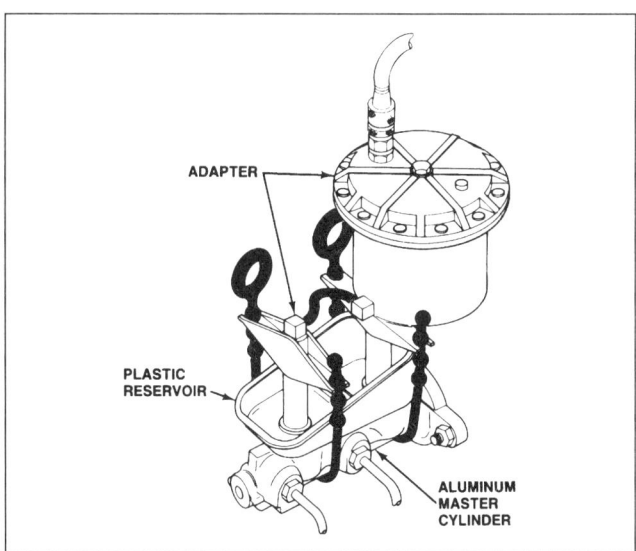

1-10 *An assortment of adapters are available for connecting a pressure bleeder to a variety of master cylinders.*

3. On vehicles with a metering or combination valve, override it with the appropriate tool.
4. Clean the top of the master cylinder, remove the reservoir cover, and clean around the gasket surface. Be careful not to allow any dirt to fall into the reservoir.
5. Select the appropriate pressure bleeder adapter and install it on the master cylinder, figure 1-10.
6. Connect the pressure bleeder fluid supply hose to the adapter, making sure the hose fitting is securely engaged.
7. Open the fluid supply valve on the pressure bleeder to allow pressurized brake fluid to enter the vehicle brake system. Check carefully for fluid leaks that can damage the vehicle finish.
8. Slip the plastic hose over the bleeder valve of the first wheel cylinder or caliper to be bled, and submerge the open end of the tube in a container partially filled with fresh brake fluid.

9. Open the bleeder valve approximately one-half turn, and let the fluid run until air bubbles no longer emerge from the tube. Then, close the bleeder valve.
10. Transfer the plastic hose to the bleeder valve of the next wheel cylinder or caliper in the bleeding sequence. Repeat steps 8 and 9. Continue around the vehicle in the specified order until the brakes at all four wheels have been bled.
11. Remove the metering valve override tool.
12. Close the fluid supply valve on the pressure bleeder.
13. Wrap the end of the fluid supply hose with a rag, then disconnect it from the master cylinder adapter. Be careful not to spill any brake fluid on the vehicle finish.
14. Remove the master cylinder adapter, adjust the fluid level to the full point, and install the fluid reservoir cover.

Vacuum bleeding

Vacuum bleeding uses a special suction pump that attaches to the bleeder valve. The pump creates a low-pressure area at the bleeder valve, which allows atmospheric pressure to force brake fluid through the system when the valve is opened, figure 1-11. Vacuum bleeding requires only one technician; however, it can only be used on wheel cylinders with cup-type expanders and brake calipers with O-ring seals. On wheel cylinders without cup expanders and calipers with stroking seals, the low pressure can pull the lips of the seals away from the bore and allow air to enter the system. To vacuum bleed a brake system:

1. Attach the open end of the vacuum pump to the bleeder valve of the first wheel cylinder or caliper

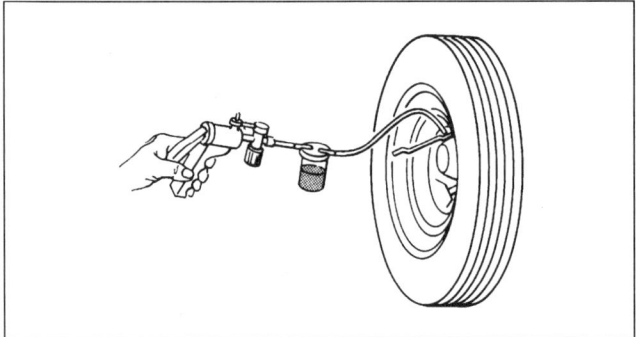

1-11 *Vacuum bleeding creates a low-pressure zone at the bleeder valve, which draws fluid through the system.*

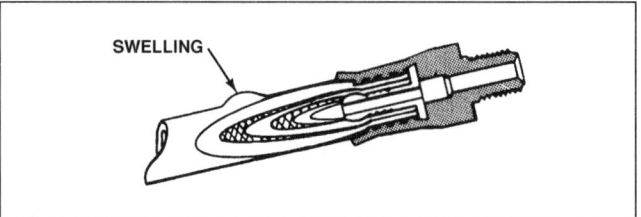

1-12 *Internal leakage, which causes a brake hose to swell or blister, is cause for replacing the hose.*

in the bleeding sequence. If necessary, use one of the adapters provided with the vacuum bleeding kit to connect to the bleeder valve.

2. Squeeze the pump handle 10 to 15 times to create a partial vacuum in the catch bottle.

3. Loosen the bleeder valve approximately one-half turn. Brake fluid and air bubbles will flow into the bottle. When the fluid flow stops, tighten the bleeder valve.

4. Repeat steps 3 and 4 until no more air bubbles emerge from the bleeder valve.

5. Transfer the vacuum bleeder to the next wheel cylinder or caliper in the bleeding sequence, and repeat steps 2 and 3. Continue in the specified order until the brakes at all four wheels have been bled.

BRAKE LINE SERVICE

Brake lines include both the rubber hoses and double-wall steel tubing that transport fluid through the system. Most manufacturers recommend that brake hoses be inspected twice a year or any time the brakes are serviced. Steel brake tubing should be inspected yearly or any time the brakes are serviced. Brake lines that are not in perfect condition must be replaced.

Brake Hose Inspection

Flexible brake hoses are used to connect steel tubing to the calipers or cylinders at the wheels. All vehicles have a hose at each front wheel to permit the wheels to move freely without damaging the brake lines. Vehicles with independent rear suspension have a hose at each rear wheel as well, while vehicles with a live, rear axle generally have a single, rear hose between the axle and the chassis.

Visually inspect brake hoses for swelling, blisters, leaks, stains, cracks, and abrasions. Swelling and blisters are signs of internal fluid leakage that has penetrated to the outer hose covering, figure 1-12. Obvious leaks or stains from leaks may appear on the surface of the hose and around the fittings on the hose

ends. Cracks can appear anywhere on the hose, as can signs of abrasion. Finally, check the hose mounting hardware and locating brackets for damage and tightness.

Swelling, which can cause poor braking performance, is not always obvious, and may only be noticeable when the brakes are applied. To check, wrap your hand around the suspect hose and have an assistant slowly depress the brake pedal. If you can feel the hose expand in your hand, it is leaking internally and must be replaced.

Brake Tubing Inspection

Visually inspect brake lines, beginning where they attach to the master cylinder, and follow them along their paths to the wheels. Look for rust and corrosion along the frame rails, at the mounting clips, or any place where water, dirt, and road salt accumulate. Also, check tubing for kinks, dents, abrasions, and other distortion and damage. Make sure the tubing is properly fastened to the chassis. Loose mounting brackets permit the tubing to vibrate, which results in cracks that lead to fluid leakage. Physical damage is most likely in the areas directly behind the wheels, or where the tubing crosses below an axle or frame member.

Brake Hose Replacement

Brake hoses that fail an inspection are replaced with a new hose. Replacement is the only common form of service because the tools necessary to fabricate new brake hoses are not readily available in the field.

Many brake hoses have a male fitting on one end and a female fitting on the other; these fittings are swaged or crimped onto the hose and do not turn. With this type of hose, the female end must be disconnected first. Other brake hoses have a banjo fitting in place of the male hose end fitting. With this type of hose, it does not matter which end is disconnected first.

When removing a hose, clean any dirt from around the fittings at the ends of the hose to prevent it from entering the hydraulic system. Hold the hose fitting securely with a wrench and use a flare-nut wrench to loosen the tubing nut, figure 1-13. Unscrew the tubing

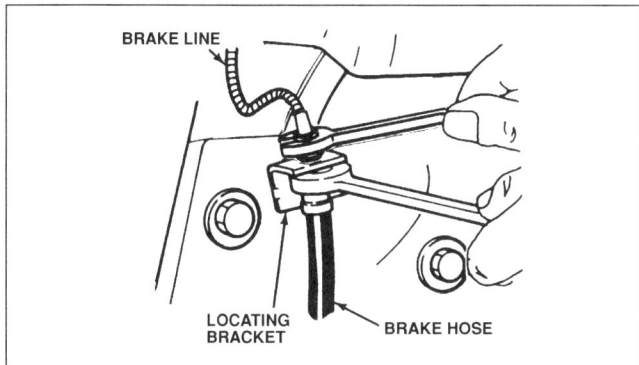

1-13 *Secure the brake hose fitting with one wrench and loosen the tubing nut with a flare-nut wrench.*

nut from the hose, remove the hose retaining clip with a pair of pliers, and separate the hose from the locating bracket, figure 1-14.

Before you install a new brake hose, make sure it is the proper part for that side of the vehicle; left- and right-side hoses are not always interchangeable. The new hose must also be the proper length, and when the original equipment hose has special armoring and bracket fittings, the replacement part must have them as well.

During installation, carefully route the new hose in the original location. Make sure the hose maintains a distance of at least 0.75 inch (20 mm) from all steering and suspension parts throughout the full range of their movement, so there is no danger of the hose being chafed. Never route brake hoses near exhaust systems, where heat will harm the rubber casing or increase brake fluid temperatures. If copper sealing gaskets are used at either of the hose fittings, use only new parts when installing the hose. Copper gaskets take a set when they are first used, and reusing an old one may cause a leak. Hold the hose fitting steady with a back-up wrench as you tighten the tubing nut to prevent the hose from twisting.

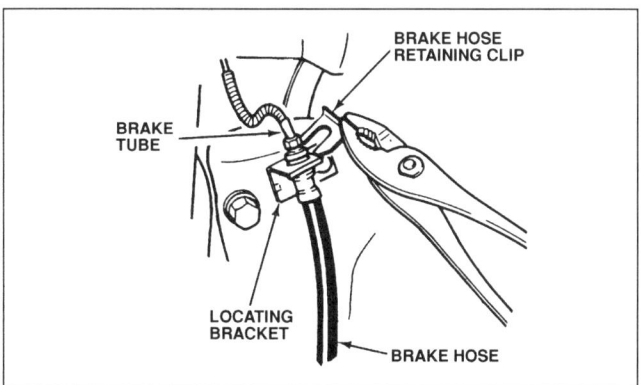

1-14 *Disconnect the tubing nut, remove the retaining clip with pliers, and separate the hose from the locating bracket.*

Brake Tubing Replacement

In most cases, custom replacement tubing that is preformed and flared is purchased for a specific application. Rarely does a technician in the field fabricate replacement tubing from raw tubing stock.

Brake tubing is held in place by clips that bolt to the chassis at various points along the line. To replace a section of tubing, you need flare-nut wrenches to loosen the fittings at both ends of the line and suitable sockets, wrenches, or screwdrivers to remove the retaining clip bolts or screws. Make sure to clean any dirt from around the tubing fittings, and use a second wrench where two brake lines connect to prevent twisting the tubing or brake hoses.

HYDRAULIC CONTROL VALVE SERVICE

Hydraulic control valve service includes procedures for testing, adjustment, and replacement. Pressure differential switches and certain height-sensing proportioning valves can be adjusted. However, neither metering valves nor proportioning valves can be repaired or adjusted. Leaking or faulty valves must be replaced. If the defective valve is part of a combination valve, the entire valve assembly must be replaced.

Recentering Pressure Differential Switches

After the brake system has been bled, the pressure differential switch may have to be recentered in order to switch off the warning light. Opening a bleeder valve creates a pressure differential between the circuits of the hydraulic system, and the switch interprets this as a fluid loss or partial system failure. In response, the piston inside the switch body moves to one side and completes the circuit to switch on the warning light.

There are three types of pressure differential switches, and each requires a different procedure to recenter it, figure 1-15. To recenter a single-piston pressure differential switch without centering springs, first determine if the brake hydraulic system is split diagonally or front to rear, figure 1-15A. Then, open a bleeder valve in the circuit of the system opposite that which was last bled, slowly depress the brake pedal until the warning light goes out, maintain pedal pressure, and close the bleeder valve.

Single-piston pressure differential switches equipped with centering springs illuminate the warning light only when the brakes are applied and a pressure difference exists between the two circuits of the brake system, figure 1-15B. The switch recenters itself automatically when the brakes are released, unless

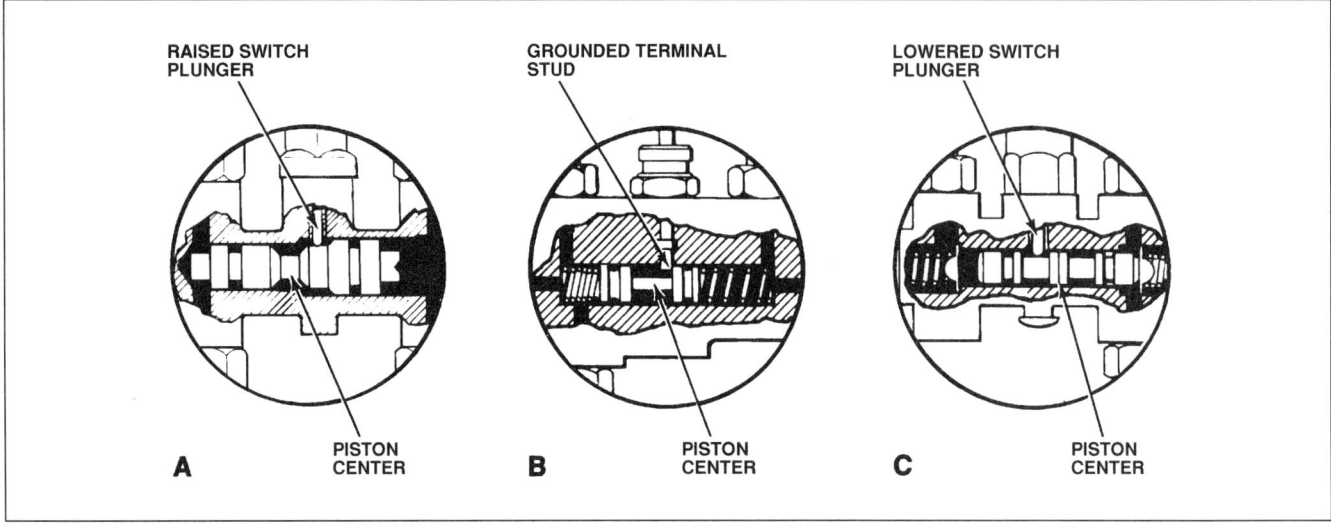

1-15 *A unique procedure is needed to recenter the piston in each of the three differential switch designs.*

the piston sticks in position against the terminal stud. If the warning light remains illuminated after the brake system has been repaired, apply the brake pedal with moderate to hard force. Hydraulic pressure should free the stuck piston and the centering springs will position it properly in the bore. The warning light will then go out.

A two-piston pressure differential switch with centering springs locks the warning light on until it is recentered, figure 1-15C. Remove the switch plunger assembly from the switch body, then apply the brake pedal with medium to hard force. This allows the centering springs to reposition the piston. Once the piston is centered, reinstall the switch plunger assembly.

Metering Valve Tests

A faulty metering valve can cause the front brakes to lock up prematurely. If you suspect a metering valve failure, begin with a visual inspection. Check around the rubber boot at the valve stem for leakage, figure 1-16. A trace of moisture is normal, but an excessive amount indicates a defective valve. Have an assistant apply the brake pedal while you watch the valve stem. As pressure to the front brakes builds, the valve stem should move. If it does not, replace the valve. More

accurate metering valve tests can be performed using pressure gauges.

Pressure gauge metering valve test

The most precise method of testing a metering valve operation is to check actual closing and opening points of the valve using a pair of pressure gauges. This test requires an assistant to apply the brake pedal, two gauges that register from zero up to a minimum of 500 psi (3450 kPa), and the appropriate fittings to attach the gauges to the hydraulic system. To perform the test:

1. Tee one of the gauges into the brake line from the master cylinder to the metering valve, figure 1-17. Make sure the gauge does not block the flow of fluid to the metering valve.
2. Connect the second gauge to one of the metering valve outlets that leads to the front brakes.

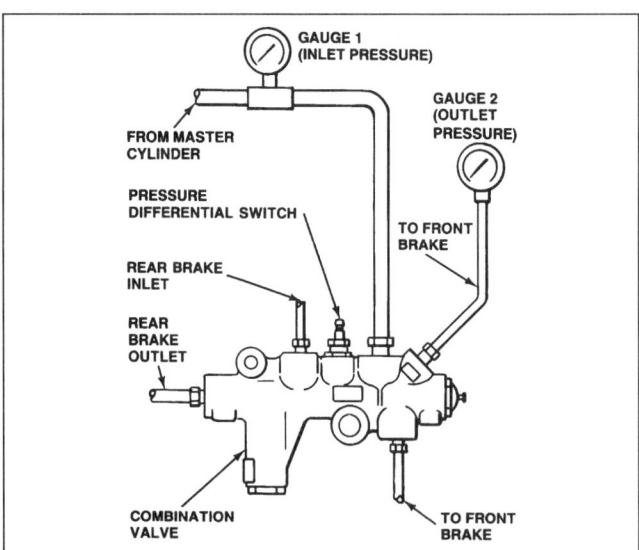

1-17 *Pressure gauge connections for testing metering valve operation.*

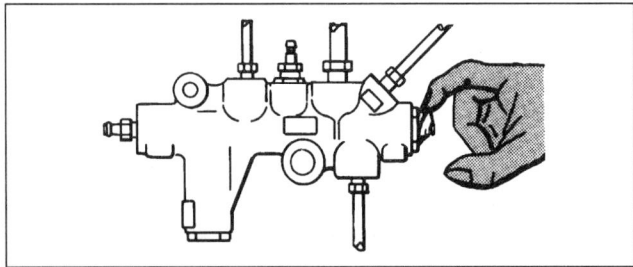

1-16 *Check behind the valve stem boot of a proportioning valve for signs of leakage.*

3. Have an assistant slowly apply the brake pedal while you observe both gauges.

4. If the metering valve is working properly, the gauge readings will rise at the same rate until they reach the valve closing point, figure 1-18A. Closing pressure, which varies by vehicle, typically falls in the 3 to 30 psi (20 to 210 kPa) range.

5. Once the metering valve closes, the reading on outlet pressure gauge (GAUGE 2) should remain constant, while the inlet pressure gauge (GAUGE 1) reading should continue to increase, figure 1-18B.

6. As inlet pressure (GAUGE 1) reaches approximately 75 to 300 psi (520 to 2070 kPa), the metering valve should open. The outlet pressure gauge (GAUGE 2) reading should then increase until it matches the inlet pressure (GAUGE 1) reading. From that point on, both gauge readings should be identical, figure 1-18C.

If the pressures indicated on the gauges do not follow the patterns described above, the metering valve is defective and must be replaced.

Proportioning Valve Tests

A typical proportioning valve failure allows rear brake pressure to increase too rapidly, which causes the rear wheels to lock prematurely during hard stops or on slippery pavement. The proportioning valve can also fail in such a way that no pressure is allowed to the rear brakes, although this is an uncommon type of failure.

Proportioning valve operation can only be tested using pressure gauges. Two pressure gauges that register from 0 to 1000 psi (0 to 6900 kPa), the appropriate fittings to attach the gauges to the hydraulic system, and an assistant are needed to test the proportioning valve. In addition, you need to know the split point of the proportioning valve on the particular make and model of vehicle being tested. Most split points are between 300 and 500 psi (2070 and 3450 kPa), but check the factory shop manual to be sure.

On vehicles where the dual braking system is split front to rear, only a single gauge hookup and test is required, figure 1-19. On vehicles with diagonal-split braking systems and dual proportioning valves, the tests are performed twice, once for each half of the hydraulic system, figure 1-20. To test the proportioning valve:

1. Tee one of the gauges (GAUGE 1) into the brake line from the master cylinder to the proportioning valve, so that the flow of brake fluid to the valve is not restricted.

2. Connect the second gauge (GAUGE 2) to the rear brake outlet of the proportioning valve.

3. Have an assistant slowly apply the brake pedal as you monitor both gauges.

4. The readings on both gauges should rise at an identical rate until the split point pressure is reached, figure 1-21A.

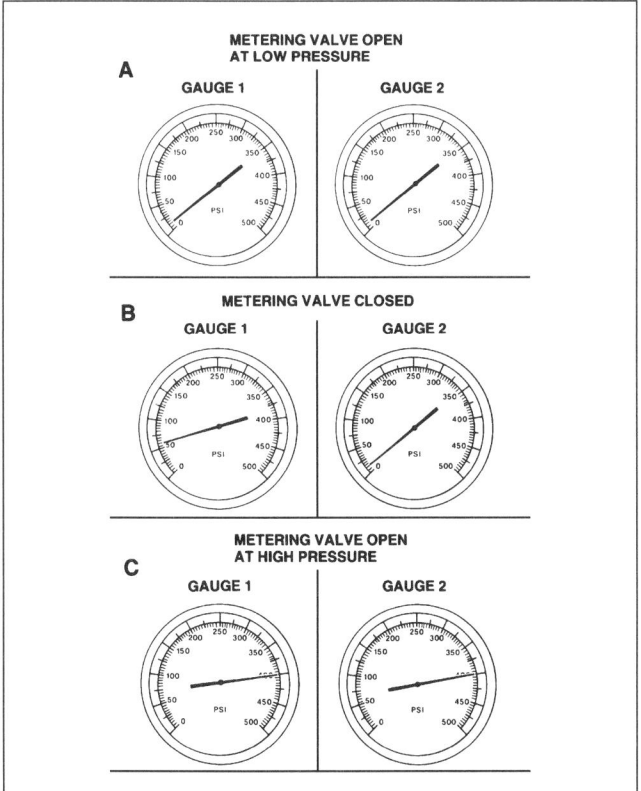

1-18 *Typical pressure gauge readings during the three phases of a metering valve test.*

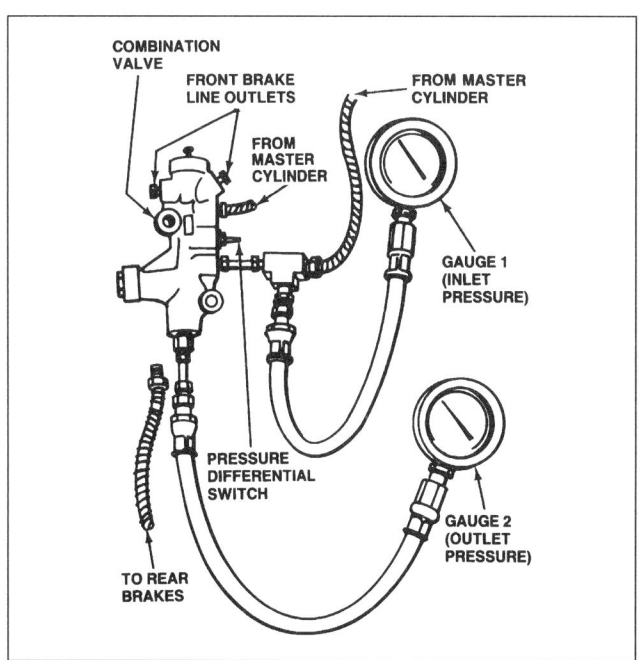

1-19 *Pressure gauge connections for testing the proportioning valve on a brake system with a front-to-rear split.*

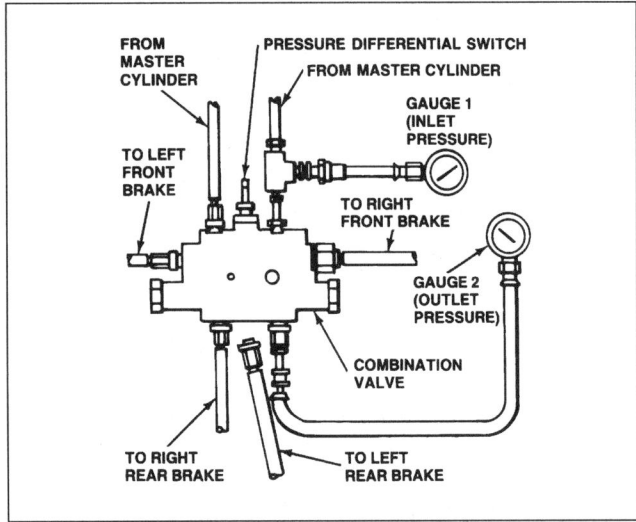

1-20 *Gauge connections for testing the proportioning valve pressure to the left, rear brake on brake system with a diagonal split.*

5. Once the split point is reached, the pressure reading on the rear brake outlet (GAUGE 2) should increase at a slower rate than the reading on the inlet pressure gauge (GAUGE 1), figure 1-21B. The proportioning valve delivers less pressure to the rear brakes than what is being produced by the master cylinder.

If the pressures indicated on the gauges do not follow the patterns described above, the proportioning valve is defective and must be replaced.

Proportioning Valve Adjustment

In some cases, a height-sensing proportioning valve must be adjusted when it is replaced. The adjustment ensures that the proportioning action takes effect at

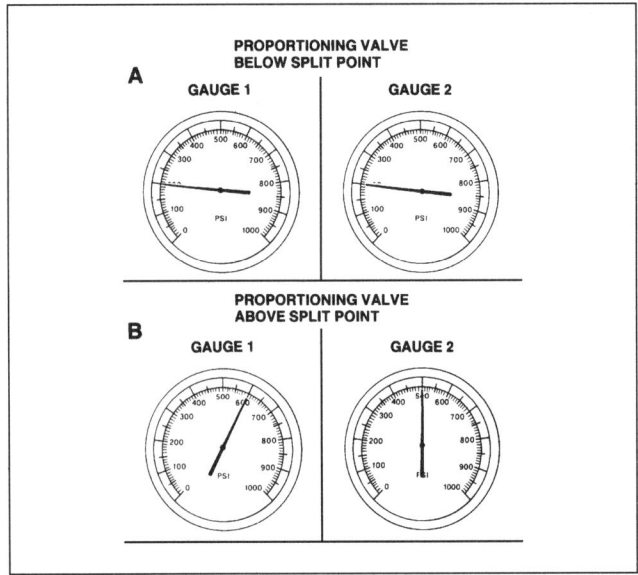

1-21 *Typical pressure gauge readings during a proportioning valve test.*

the correct hydraulic pressure in relation to vehicle loading. There are nearly as many adjustment procedures as there are variable proportioning valves. The adjustment procedures are given below for two types of valves in use today. Consult the factory shop manual for the exact procedure on other types of valves.

The height-sensing proportioning valve on some trucks requires a special plastic adjustment gauge. This gauge, which is a one-time use item that is supplied with the replacement part, installs on the valve to hold it in position during installation, figure 1-22. Once the valve operating lever is tightened in place, a tang on the gauge is cut away to allow unrestricted valve operation, figure 1-23.

Other height-sensing proportioning valves require you to set a spring length, figure 1-24. Adjust the distance by loosening the spring support bolt and sliding the spring support until the distance is correct.

STOPLIGHT CIRCUIT TESTS AND ADJUSTMENTS

There are two potential problems in the stoplight circuit: Either the stoplights are always on, or they fail to come on when the brakes are applied. Most of these problems result from burned out fuses or bulbs, a faulty stoplight switch, faulty wiring, or a switch that is out of adjustment.

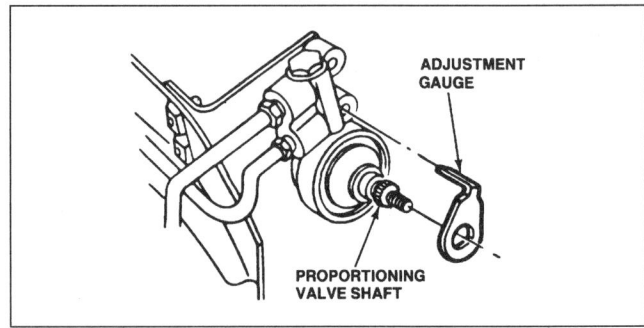

1-22 *This disposable plastic adjustment gauge holds the proportioning valve in position during installation.*

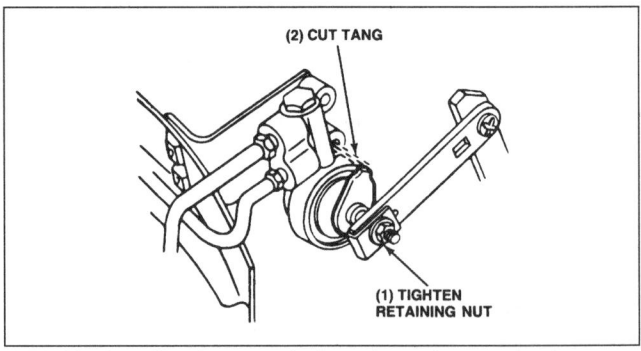

1-23 *After installing the proportioning valve and connecting the operating lever, cut the positioning tang off the plastic adjustment gauge to release the valve.*

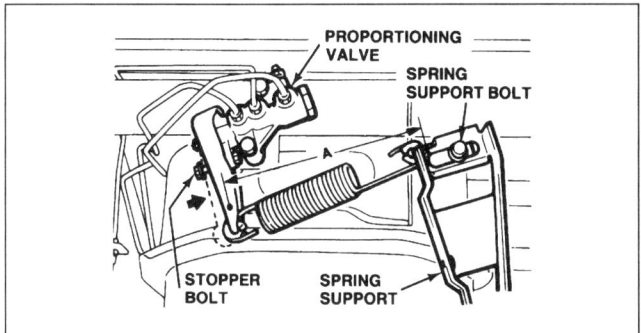

1-24 *This height-sensing proportioning valve adjusts by loosening the spring support bolt and positioning the support to achieve a specific spring length.*

Virtually all late-model vehicles have mechanical switches on the brake pedal linkage. Mechanical switches require adjustment when they are installed.

Mechanical Stoplight Switch Adjustment

There are two basic methods to adjust stoplight switches. After performing either adjustment, check the brakes lights for proper operation.

Stoplight switches for most import vehicles have a threaded shank that screws into place and is secured in position by a locknut, figure 1-25. To adjust these switches, loosen the locknut on the threaded shank of the stoplight switch body. Turn the switch to move it in or out of its mounting bracket and obtain the correct clearance between the brake pedal arm and the switch plunger. When properly adjusted, depressing the brake pedal 0.5 inch (13 mm) will illuminate the brake lights.

Many domestic vehicles have a semi-automatic adjustment mechanism, figure 1-26. To install and adjust, depress the brake pedal and insert the switch to fully

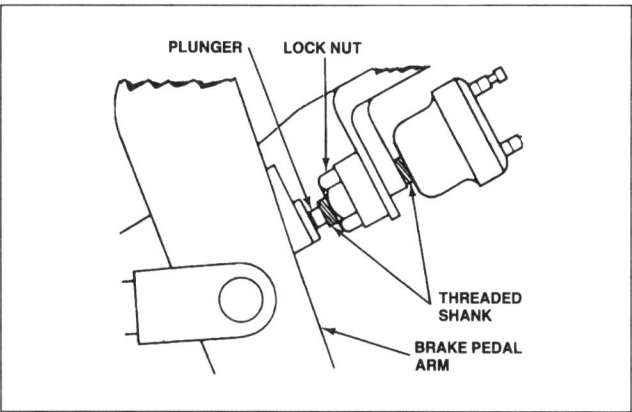

1-25 *To adjust a stoplight switch with a threaded shank, loosen the locknut, thread the switch in or out to the desired length, then tighten the locknut.*

seat it into the tubular clip on the brake pedal mounting bracket. The switch clicks into place when fully seated. Pull up on the brake pedal to adjust the switch position. This causes the switch to ratchet in the tubular clip and emits a series of clicks as it does so. Release the pedal, and repeat the previous step until the clicking of the ratchet stops.

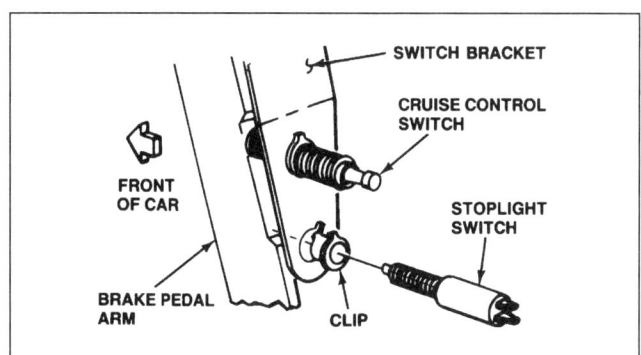

1-26 *This stoplight switch installs with a tubular clip that semiautomatically adjusts the switch position.*

1. Brake pedal reserve that gradually fades under light pressure indicates:
   a. A brake line restriction
   b. Spring tension on the brake pedal is weak
   c. Internal leaks
   d. The system is operating properly

2. Which of the following statements is NOT true?
   a. Some imports use HSMO, a green-colored oil, as brake fluid
   b. Silicone, or DOT 5 brake fluids, contain a purple dye
   c. Polyglycol brake fluids are amber to clear in color
   d. All DOT-approved brake fluids are hygroscopic

3. A bypassing master cylinder can be detected by the:
   a. Internal leak test
   b. External leak test
   c. Compensating port test
   d. QTU test

4. Honing will remove the anodized finish on which type of master cylinder?
   a. Add-on ABS
   b. Aluminum
   c. Cast iron
   d. Composite

5. The first piston installed when rebuilding a master cylinder is the:
   a. Primary piston
   b. Secondary piston
   c. QTU piston
   d. Compensating piston

6. The QTU valve on a master cylinder regulates fluid flow:
   a. Between the cylinder and the reservoir
   b. Between the two chambers of the reservoir
   c. Between opposite wheels
   d. Between the two channels of the master cylinder

7. The master cylinder should be cleaned with:
   a. Kerosene or carburetor cleaner
   b. Brake parts cleaner or brake fluid
   c. Fresh solvent or hot, soapy water
   d. Motor oil or cutting oil

8. A correctly honed cylinder bore should:
   a. Have a smooth, polished surface
   b. Have a cross-hatch pattern
   c. Be larger than the original size by a specified amount
   d. Have scoring no deeper than 0.006 inch (0.15 mm)

9. The first component in a brake system bleeding sequence is the:
   a. Master cylinder
   b. Wheel cylinder
   c. Combination valve
   d. Proportioning valve

10. Generally speaking, which of the wheel brakes should you bleed first?
    a. The one farthest from the master cylinder
    b. The one closest to the master cylinder
    c. The one closest to the driver
    d. The one closest to the passenger

11. What special tool is required to pressure bleed a system equipped with a metering valve?
    a. An adjustment gauge
    b. A torque wrench
    c. A vacuum tool
    d. An override tool

12. On a front brake hose that does not have a banjo fitting, which end do you disconnect first?
    a. Female end
    b. Male end
    c. Caliper end
    d. Either end

13. After bleeding brakes:
    a. The pressure differential switch may need to be recentered
    b. The metering valve may need to be adjusted
    c. The proportioning valve may need to be recentered
    d. The pressure differential switch may need to be adjusted

14. The most precise way to test metering valves is with:
    a. A pressure bleeder
    b. Pressure gauges
    c. An override tool
    d. A vacuum gauge

15. Height-sensing proportioning valves:
    a. Cannot be adjusted
    b. Are used only on FWD vehicles
    c. Can be adjusted in many cases
    d. Are often the cause of brake problems

Chapter Two

DRUM BRAKE SERVICE

Begin brake drum diagnosis with a road test. When applying the brakes, listen for unusual noises, pay attention to how the brake pedal feels as the brakes activate, and note if the vehicle pulls as it comes to a stop.

Excessive pedal travel before the brakes apply is often the result of incorrect adjustment; the shoes must move too far before contacting the drum. A vehicle that pulls to one side is the result of uneven shoe-to-drum clearance from side to side, or a shoe that is hanging up or binding. Pulling due to a drum brake problem is subtle on a vehicle with front disc and rear drum brakes, and is often undetectable without removing the drums and inspecting the brake assembly.

DRUM BRAKE NOISE

Listen carefully when applying the brakes during a road test; noises can reveal a number of drum brake problems. A scraping or grinding sound usually indicates worn out brake linings that have resulted in metal-to-metal contact between the shoe lining table and drum. A thumping sound when the brakes apply is usually caused by a cracked brake drum, defective brake shoes, or weak brake shoe return springs. The exact cause can be determined during an inspection. A loud scraping noise that occurs as the brakes are released can be caused by a bent backing plate contacting a brake drum.

BRAKE DRUM REMOVAL

To service a drum brake, you must first remove the drum to gain access to the friction assembly. The removal procedures differ for **fixed** and **floating drums**. However, both designs require loosening the parking and service brake adjustments before the drum can be removed. This is necessary because wear at the open edge of the drum, or scoring of the brake linings and

drum friction surface, creates a ridge or several interlocking grooves that hold the drum in place, figure 2-1. Always mark brake drums before removing them; they must be returned to the same axle and the same side of the vehicle on assembly.

Fixed Drum Removal

Fixed brake drums are common on the rear axles of front-wheel drive (FWD) vehicles and the front axles of older rear-wheel drive (RWD) vehicles with front drum brakes. In most applications, the nut that retains a fixed brake drum also secures and preloads the wheel bearings in the hub.

To remove a fixed drum, remove the dust cap from the center of the hub. Most designs have either a castellated nut or a castellated retainer that fits over a standard nut. Both are secured by a cotter pin. Remove the cotter pin along with any other locking devices, then remove the nut. Pull outward on the drum to slide it off the spindle. Take care not to let the thrust washer and outer wheel bearing fall on the ground as they clear the spindle. Also, avoid dragging the inner wheel bearing across the retaining nut threads on the spindle.

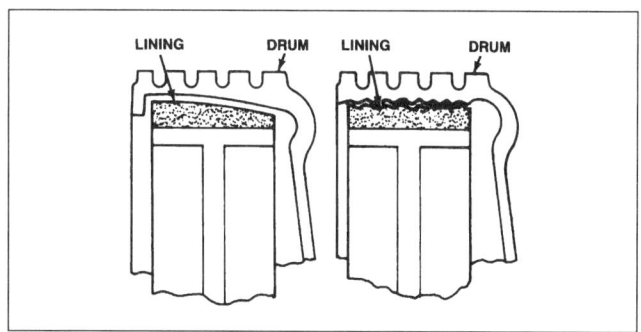

2-1 Brake lining and friction surface wear make it difficult to remove a drum unless the parking and service brake adjustments are loosened.

Fixed Drum: A brake drum that is cast in one piece along with the hub assembly, which contains the wheel bearings. Fixed drums are most often used on non-driven axles, such as at the rear of a front-wheel drive vehicle.

Floating Drum: A brake drum that installs on a separate axle flange or hub assembly. Floating drums, which are commonly found on the drive axle of rear-wheel drive vehicles, can also be used on a non-driven axle.

Once the drum is removed, inspect the grease in the hub and on the wheel bearings. If the grease is relatively fresh and in good condition, set the drum on the bench, open side down, and cover the outer bearing and hub opening so the bearings will not become contaminated. If the grease is old and dirty, repack the wheel bearings before you reinstall the drum. Always repack the wheel bearings when performing a complete brake job.

Floating Drum Removal

Floating brake drums are held in place by the wheel and lug nuts during normal operation. However, some designs have small bolts or screws that fit through a hole on the face of the drum and into a threaded hole on the axle flange. These must be removed before the drum can be removed. Some drums may also be retained by **speed nuts**, which fit over the wheel studs that protrude through the drum from the hub or axle flange. To remove speed nuts, grasp them with a pair of pliers and thread them off the studs, figure 2-2. Once removed, speed nuts can be discarded because they are unnecessary as soon as the vehicle leaves the factory. Now, the drum should move freely on the hub or axle and slip off over the brake shoes.

BRAKE DRUM INSPECTION

Two inspections are made on a brake drum, a visual inspection followed by one or more careful measurements. After completing a thorough inspection, you will know if the drum or rotor is in serviceable condition or is beyond saving and must be replaced. A serviceable drum can be machined to restore the friction surface, then reinstalled to its original position on the vehicle.

Visual Inspection

To inspect a drum, first wipe its friction surface clean with a shop towel soaked in brake parts cleaner to make it easier to spot any problems. During the visual inspection, if any problems that require the drum to be machined are found, immediately measure the drum as described later in the chapter. Visually inspect the drum for:

- Scores and grooves
- Cracks
- Heat checks
- Hard spots.

Scores and grooves on the drum friction surface increase brake wear and noise. To determine the depth of any scores or grooves, use a micrometer with a pointed anvil designed for this purpose. As a general rule, any score or groove deeper than 0.010 inch (0.25 mm) requires turning the drum.

Cracks can occur anywhere, but drums usually crack near the bolt circle or web and at the open edge of the friction surface. Do not confuse small surface cracks with cracks that reach deeply into the structure of the drum. If any deep cracks are visible, replace the drum.

Heat checking appears as many small, interlaced cracks on the friction surface, figure 2-3. Heat checks can cause a slight pedal pulsation, increase brake lining wear, and make noise. If the heat checking is minor and the drum checks out in other respects, machine the drum to repair. If heat checking is widespread, replace the drum.

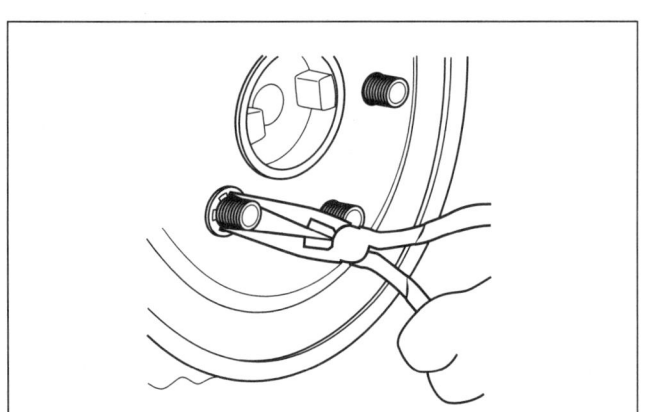

2-2 *Speed nuts, which hold a floating drum in place on the assembly line, can be removed and discarded during service.*

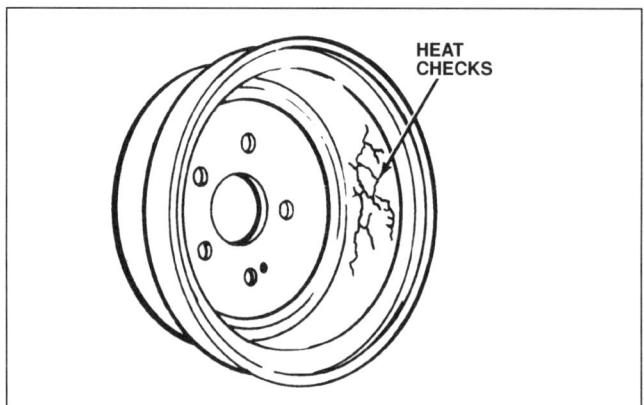

2-3 *Heat checks, which are a series of small, interlaced cracks on the friction surface of the drum, can be machined out if they are minor and localized.*

Speed Nut: A spring-steel clip that threads onto a stud or bolt to position a part. Speed nuts are used to hold floating drums and rotors in place during vehicle assembly.

Hard spots are round, bluish/gold, glassy appearing areas that develop on the friction surface, figure 2-4. Hard spots can cause pedal pulsation, brake chatter, and increase lining wear. Since machining down hard spots requires special equipment and is time-consuming, common practice is to replace drums that have hard spots.

Measurement

Brake drums are measured to identify wear and distortion that is not visually apparent. When drums wear, they become oversized, tapered, or barrel-shaped. Distorted drums become bell-mouthed, out-of-round, or eccentric. Most of these problems can be detected by using either a drum micrometer or an inside micrometer. However, some forms of drum wear and distortion cannot be identified until the drum is actually turned in a brake lathe.

Drum inside diameter

Anytime a brake drum is removed from the vehicle, inside diameter should be measured to check for wear. To measure, first note the discard diameter stamped or cast into the drum, figure 2-5. Then, position the drum so the open side is facing up.

Adjust a drum micrometer to the nominal drum diameter, fit it into the drum, and take measurements at two or three locations, figure 2-6. Compare the largest micrometer reading to the discard diameter stamped on the drum.

As a general rule, if the inside diameter is not at least 0.030 inch (0.75 mm) smaller than the drum discard diameter, replace the drum. The amount of additional metal required to allow for wear in service varies by manufacturer; check the shop manual of the vehicle being serviced for the exact value. If the drum

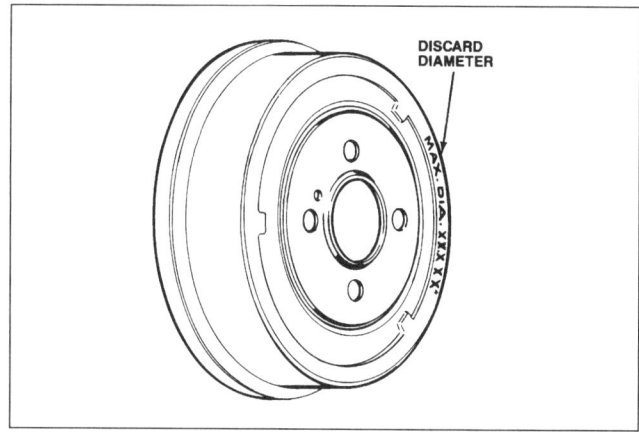

2-5 *Discard diameter, which is stamped or cast into the face of the drum, shows the machining and wear limits of the drum.*

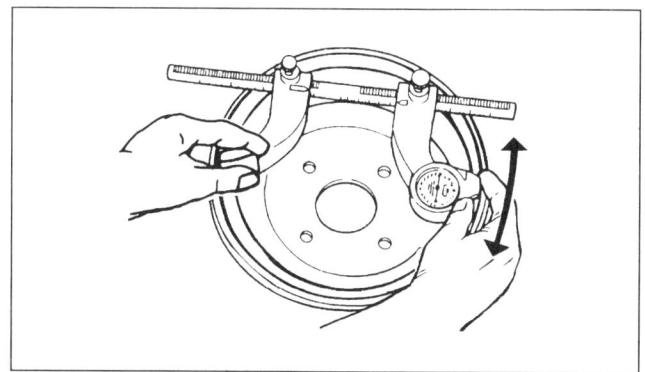

2-6 *Use a drum micrometer to take diameter measurements at several locations around the drum.*

needs to be turned, there must be sufficient metal remaining, so that the inside diameter remains at least 0.030 inch (0.75 mm) smaller than the discard diameter after machining.

Ford Motor Company drums, as well as those used by some other manufacturers, are marked differently. Replace one of these drums whenever the inside diameter measurement exceeds the maximum diameter stamped or cast into the drum. When unsure of what the measurements on a drum mean, consult a shop manual to be sure.

Drum taper wear, barrel wear, and bellmouth distortion

Taper wear, barrel wear, and bellmouth distortion are problems that cause variations in brake drum diameter between the open and closed edges of the friction surface. A drum with taper wear has a larger diameter at the closed edge than at the open edge, while a drum with barrel wear has a larger diameter at the center than at either edge, figure 2-7. A drum with bellmouth distortion has a larger diameter at the open edge than at the closed edge, figure 2-8.

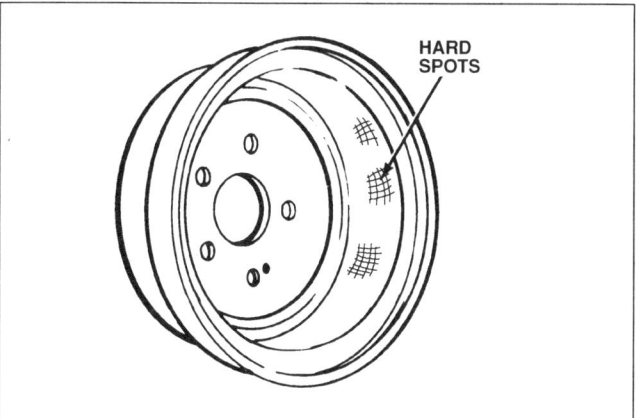

2-4 *Hard spots, which appear where extreme heat has hardened the iron in the drum, can cause pedal pulsation, chatter, and increase lining wear.*

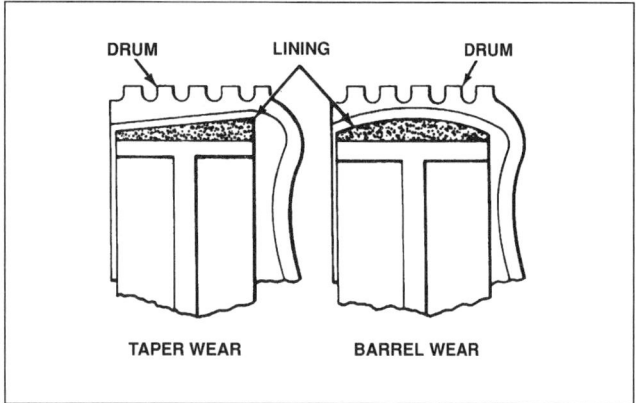

2-7 *Taper wear, the largest, inside diameter at the closed edge of the friction surface, and barrel wear, the largest, inside diameter at the center of the friction surface, are common in brake drums.*

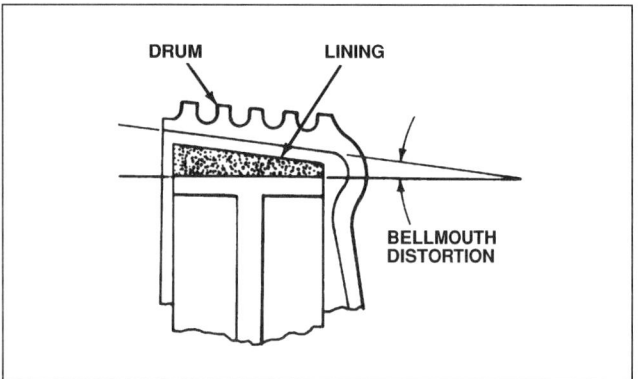

2-8 *Bellmouth wear, the largest, inside diameter at the open edge of the friction surface, occurs less frequently than other types of brake drum wear patterns.*

Taper wear can sometimes cause a spongy brake pedal, but barrel wear and bellmouth distortion have no symptoms that are obvious to the driver. These problems can sometimes be spotted by ridges or lips worn into the drum friction surface; other times, unusual wear patterns on the brake linings will reveal the problem.

You can also identify these wear patterns by measuring the drum inside diameter at several points across the friction surface. A drum micrometer cannot reach deeply enough into the drum to make these measurements, so you must use an inside micrometer instead.

Position the inside micrometer in the drum and take three measurements, one at the open edge of the drum, one at the center of the friction surface, and one at the closed edge of the drum, figure 2-9. If the highest and lowest of these measurements vary by more than 0.006 inch (0.15 mm), machine the drum. Replace the drum if machining will not leave the inside diameter at least 0.030 inch (0.75 mm) smaller than the discard diameter.

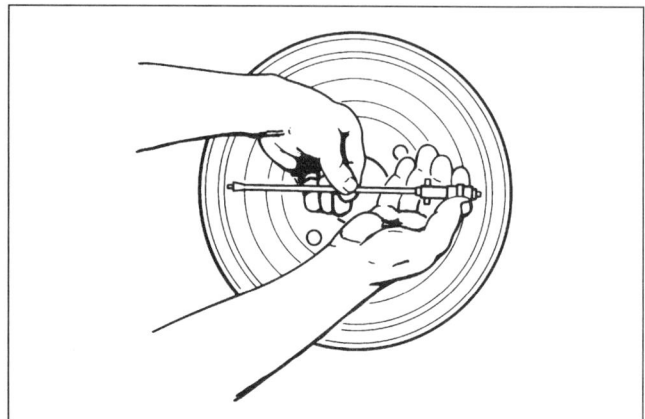

2-9 *Take three inside micrometer readings, at the closed edge, center, and open edge of the friction surface, to check a drum for taper, barrel, and bellmouth wear.*

Out-of-Round Drum Distortion

The diameter of an out-of-round drum varies when measured at several points around its circumference. This causes a pulsating brake pedal, brake vibration, and sometimes, grabby, erratic braking. To check for an out-of-round drum, use a brake drum micrometer to measure the drum inside diameter at four locations 45 degrees apart from one another, figure 2-10. If the highest and lowest measurements vary by more than 0.006 inch (0.15 mm), correct by machining the drum. Remember, final diameter must be 0.030 inch (0.75 mm) smaller than the discard diameter.

Eccentric Drum Distortion

Eccentric brake drum distortion exists when the geometric center of the friction surface is different from that of the hub. This makes the drum rotate with a cam-like motion, which causes the shoe contact pads on the backing plate to wear and creates noise whenever the brakes are applied. Since eccentric drum distortion does not affect inside diameter, it cannot be detected visually or with common measuring tools.

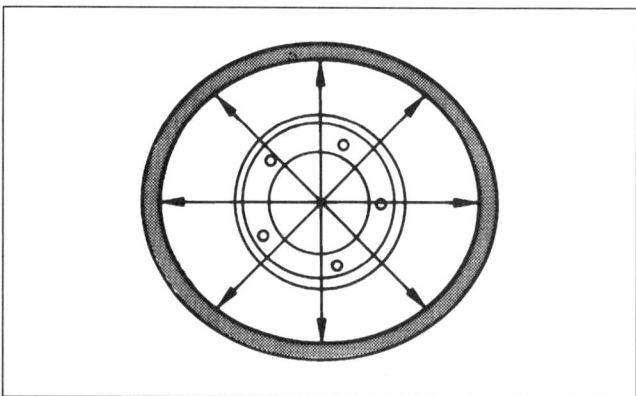

2-10 *Take four micrometer readings at 45 -degree increments around the circumference to check a brake drum for out-of-round distortion.*

This condition is identified while a drum is being turned on a lathe and the tool bit contacts the friction surface on only one side of the drum.

BRAKE DRUM MACHINING

Brake drum **turning** uses a brake lathe and a steel tool bit to remove metal from the friction surface of the drum, figure 2-11. Turning can repair most forms of wear, damage, and distortion. When turning a drum, remove only the minimum amount of metal necessary to restore the friction surface. This helps ensure the longest possible service life for the drum.

Never machine only one drum. Always machine both drums on each end of the same axle to the same inside diameter. This keeps braking force and fade resistance equal from side to side and prevents brake pull. The inside diameters of drums on the same axle should be kept within 0.020 (0.51 mm) of each other. To help keep the drum diameters equal, machine the more badly worn drum first, then machine the other drum to match. In the same manner, machine a new drum to match the diameter of an old drum on the same axle when only one drum is being replaced.

BRAKE SHOE REPLACEMENT

The thickness of the lining material is the main factor that determines whether brake shoes should be replaced. The brake shoe friction linings must be at least 0.030 inch (0.75 mm) above the lining table or rivet heads, figure 2-12.

Brake shoes are sold and serviced as axle sets. An axle set consists of four shoes; one pair for the friction assembly at each wheel. Shoes from different manufacturers should never be mixed. Although they will fit and may appear the same, the friction coefficients of

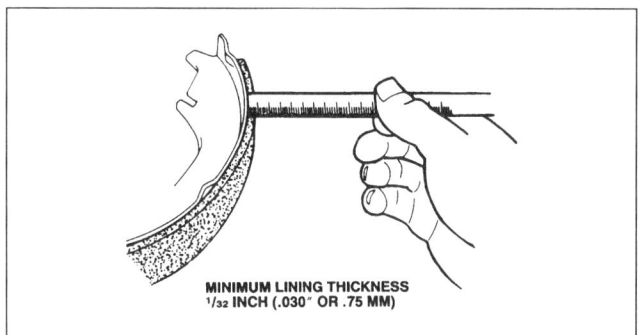

2-12 *Measure the thickness of the brake shoe friction lining with a machinist scale to determine wear.*

the linings may be quite different. Even if only one shoe of an axle set is badly worn, the entire set should be replaced after the problem causing uneven wear has been repaired.

The exact procedure used to replace a set of brake shoes varies with the design of the friction assembly and the hardware used to mount and activate shoes. The following sections describe a general procedure that includes the common operations required to disassemble, inspect, and reassemble drum brakes.

Drum Brake Disassembly

The order in which the many parts of the friction assembly are disassembled varies from one brake to another. However, once the drum is removed, removing the shoe return springs is often the next step. Avoid injury by always wearing eye protection when removing and installing springs.

Springs that hook over an anchor post are removed with a special brake spring tool, figure 2-13. Place the tool over the post and hook the flange under the end

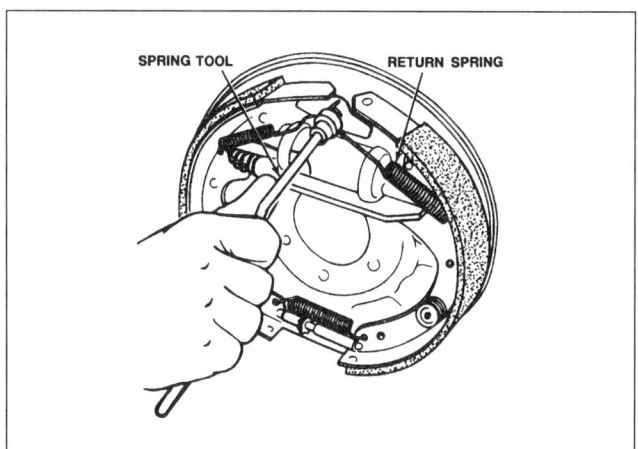

2-13 *A special brake spring tool is used to lever return springs off the anchor post.*

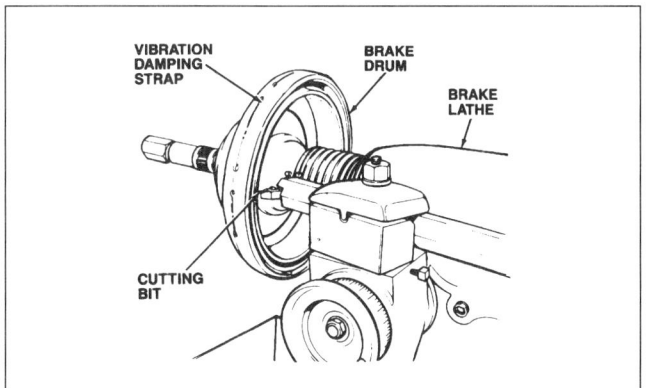

2-11 *The turning process uses a tool-steel cutting bit to remove metal from the friction surface as the drum rotates on a brake lathe.*

Turning: A machining process that uses a brake lathe to remove metal from drums and rotors to refinish their friction surfaces.

of the spring. Rotate the tool to lever the spring up and off the anchor. Once the springs are free, remove the anchor plate and adjuster cable or linkage, if fitted, from the anchor post.

Return springs that install between two brake shoes can be removed with a pair of brake spring pliers. However, this is often unnecessary because once the shoes are removed from the anchor, they can be collapsed together by hand to release the tension on the shoe-to-shoe spring.

Once the springs are off, remove the brake shoe holddowns. Although most holddown devices can be removed by hand or with common hand tools, there are special tools available that make the job easier, figure 2-14.

After removing the return springs and shoe holddowns, the friction assembly can be lifted free of the backing plate. In most cases, simply reposition the brake shoes as needed so the shoe-to-shoe return springs, automatic adjusting mechanism, and parking brake linkage can be disconnected.

Drum Brake Inspection

Inspect the brake backing plate and its mounting bolts. If the plate is bent or cracked, replace it. If the mounting bolts are loose, tighten them to the torque specification recommended by the manufacturer. Also, check the shoe support pads for grooves, notches, or any other signs of wear. Although minor wear is considered normal, smooth the pads by filing or grinding to provide a good surface for the new shoes to ride on. If the pads are deeply grooved, replace the backing plate. Then, inspect the wheel cylinder as described in the next section.

Most vehicle manufacturers recommend that the shoe return springs, shoe holddown hardware, and automatic adjuster cables be replaced whenever a new set of brake shoes are installed. This is because it is difficult to determine the condition of these parts by inspection.

The automatic adjusting mechanism and parking brake linkage can generally be reused if they pass a visual inspection. To inspect, look for bent components and wear at the points where the parts contact one another. If the brake assembly has starwheel adjusters, disassemble the adjuster assembly and clean it thoroughly, figure 2-15. Replace the adjuster if the starwheel teeth are rounded, chipped, or broken. Use a wire brush to clean the adjuster threads, lubricate the threads with brake grease, then assemble the adjuster and thread it through its full range of travel. If the threads bind at any point, repair the problem or replace the adjuster.

Wheel cylinder inspection

Grasp the wheel cylinder and attempt to move it. If any movement is detected, make sure all of the mounting hardware is in place and properly tightened. Some import vehicles have wheel cylinders that slide in a slot on the backing plate. These cylinders are designed to move, so simply make sure the mounting clips are present and properly installed. The dust boot that seals the slot must be in good condition.

Inspect the outside of the wheel cylinder for leaks. Minor stains caused by fluid seepage are considered normal. Fold back the cylinder dust boots and look for liquid. If you find more than a slight amount of dampness, rebuild or replace the cylinder.

Next, check for free movement of the wheel cylinder pistons. With the brake drum from only a single wheel removed, have an assistant gently apply and release the brake pedal while you verify that both brake shoes move outward and return smoothly to their stops. On brakes without piston stops, use two large screwdrivers to make sure the pistons are not pushed out of the cylinder bore. Insert the tips of the screwdrivers under the lip at the edge of the backing plate, then lever the screwdriver shafts against the brake shoes to prevent them from moving outward too far.

On some brake designs, a frozen wheel cylinder piston can prevent one or both of the brake shoes from

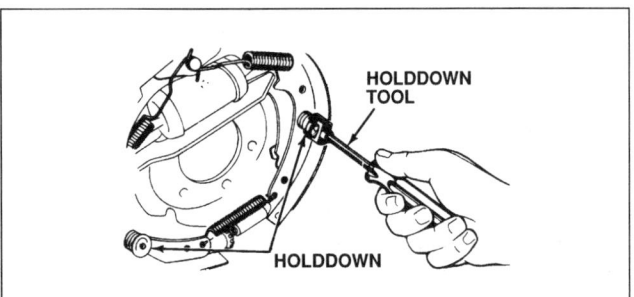

2-14 *Special tools make removing spring-and-pin brake shoe holddowns easy.*

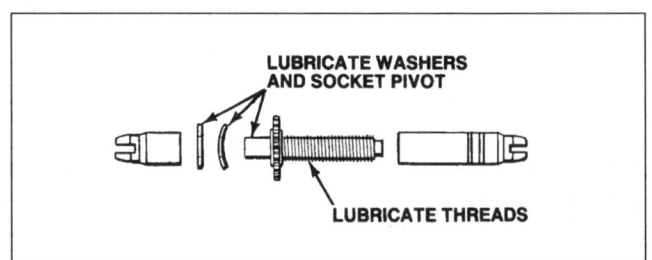

2-15 *Disassemble, clean, and inspect starwheel adjusters. If they pass inspection, they can be lubricated, assembled, and reinstalled.*

applying. A sticking piston that slows or prevents full return of the brake shoes will cause the brakes to drag, resulting in rapid lining wear and possibly brake fade.

Wheel cylinder service

Many wheel cylinders can be taken apart, internally inspected, and rebuilt while they are still mounted on the backing plate. Disassemble a wheel cylinder simply by removing the shoe links and the rubber boots from the ends of the cylinder. Then, press out the pistons, cup seals, spring, and cup expanders, figure 2-16. Inspect, hone, and measure the bore of a wheel cylinder using the same techniques detailed in chapter one for servicing a master cylinder.

Brake Assembly

Drum brake assembly is essentially the reverse of disassembly, making sure that all parts are reinstalled in their proper locations. There are also a few special techniques used to install specific brake components.

To begin, compare the replacement brake shoes to the original equipment parts. They should have the appropriate holes in the **shoe webs**, and the linings should be the same basic size and shape as those on the originals.

Next, determine where the shoes belong on the vehicle. For example, the **primary shoe** on a **dual-servo brake** generally has a smaller lining than the **secondary shoe**, and the friction materials used for the two linings may differ as well. Always install the

primary shoe so it is pulled away from the anchor when the brakes are applied, with the wheel turning in the direction of forward rotation. Some **leading-trailing brakes** are also designed to use shoes that have different friction characteristics. In these applications, install the replacement shoes in the same relative locations as the original equipment parts.

Remove any parking brake linkage pieces or similar parts from the old shoes. Transfer these parts to the appropriate replacement shoes and install them using new fasteners. Lubricate the shoe support pads on the backing plate with a thin coat of high-temperature brake grease, figure 2-17. Avoid excess lubrication, as grease liquifies at high temperatures. If too much grease is applied, it can run and be absorbed by the friction material, which results in braking problems.

Assemble the shoes onto the backing plate. The procedure varies with the design of the friction assembly; basically, reposition the shoes as needed until the parking brake linkage, brake adjuster, and the shoes themselves are all fitted together in their proper positions.

Once the shoes are assembled in position, install the holddowns to secure the shoes in place. On pin and spring holddowns, insert the pin through the holes in the backing plate and brake shoe web. Then, while holding the pin in place from the backside of the backing plate, use the special tool to compress the spring and retaining washer over the end of the pin. Rotate the washer as needed to lock it onto the flattened end

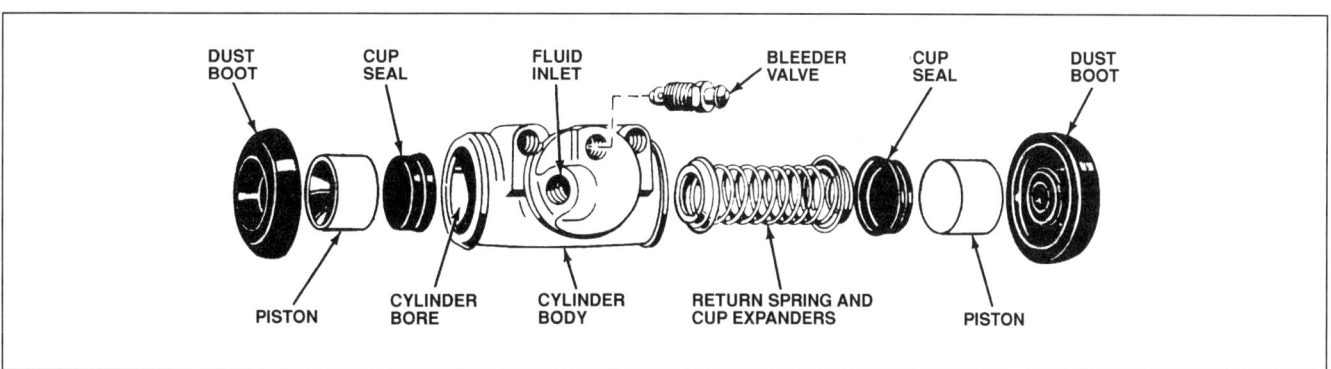

2-16 *Most wheel cylinders can be disassembled, inspected, and serviced without removing them from the vehicle.*

Shoe Web: The portion of the brake shoe below the lining table that receives the application force from the wheel cylinder.

Primary Shoe: The shoe in a servo brake that transfers a portion of its stopping power to the secondary shoe.

Dual-servo Brake: A drum brake that has servo action, or force transmitted by the wheel cylinder, in both the forward and reverse directions. The rear drum brakes on most rear-wheel drive domestic vehicles are dual servo.

Secondary Shoe: The shoe in a servo brake that receives extra application force from the primary shoe. The lining of a secondary shoe is larger than that of the primary shoe because it does most of the braking.

Leading-trailing Brake: A non-servo brake with one leading shoe and one trailing shoe.

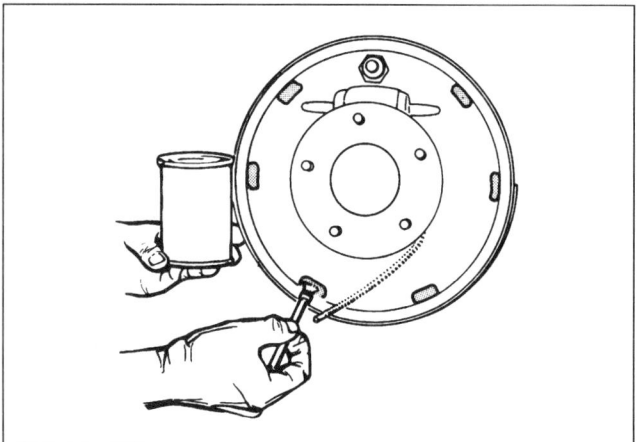

2-17 *Apply a thin coat of high-temperature brake grease to the shoe contact pads on the backing plate.*

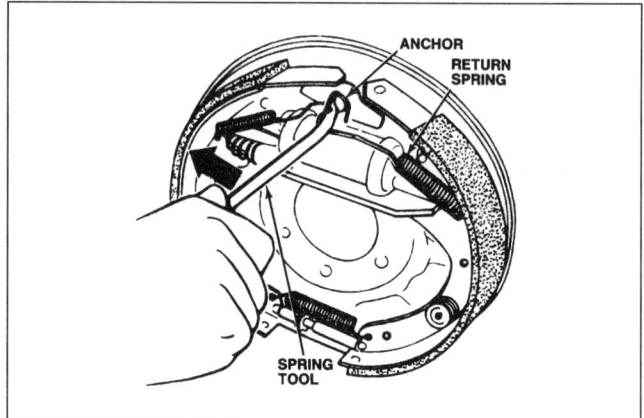

2-18 *Using a brake spring tool to install a return spring on an anchor post.*

of the pin. Where a spring clip is used with a holddown pin, compress the clip by hand and slip it into position under the flattened end of the pin. To install a coil-spring beehive holddown, hold the retaining clip in place from the backside of the backing plate, then use a Phillips screwdriver to push the spring inward, engaging its hook into the retaining clip.

The final step in drum brake assembly is to install the shoe return springs. It is important to install the return springs facing in the proper direction and in the correct location. Some springs can be installed only one way, and their proper position is easy to identify. Different paint colors are often used to distinguish similar springs that have different tensions. Certain springs have a longer straight section and attachment hook at one end than at the other. These springs must be installed facing a specific direction, or the coiled section of the spring will interfere with another part of the friction assembly. Sometimes, there are several holes in the shoe web where a spring can be attached. If you install a spring in the wrong hole, it will affect the rate at which the brake shoes apply and release.

Return springs are installed in two basic ways, and both methods require a special tool. To install a spring that fits over an anchor post, attach the appropriate end of the spring into the hole in the shoe web. Then, place the notched end of the spring tool on the anchor post, and drape the hook end of the spring over the tool shaft, figure 2-18. Take care not to overstretch the spring as you lever the tool back, so the spring slides down the shaft and into place on the anchor.

Shoe-to-shoe return springs are installed using brake spring pliers, figure 2-19. Fit one end of the spring into the correct hole in the brake shoe web and place the other end of the spring over the hooked arm of the brake spring pliers. Position the pliers over the shoe the spring is to be attached to so that the

pointed arm of the plier contacts the lining at the same level as the hole the spring is to engage. If possible, position the pointed arm on a lining rivet. Otherwise, position it directly on the lining and use extra caution to prevent damage. Squeeze the handle of the pliers to stretch the spring to the appropriate length, then insert the end of the spring into the hole in the web. Remove the plier, and make sure both ends of the spring are fully engaged in the web holes.

Initial Brake Adjustment

Once the brake is assembled, adjust the initial lining-to-drum clearance so the brake pedal travel will be satisfactory. With starwheel brake adjusters, initial adjustment is done before the drum is installed.

Initial Adjustment Starwheel Adjusters

Both manual and automatic starwheel brake adjusters make their adjustments in very small increments. If you attempt to adjust the brakes manually after the drum is installed, a considerable number of adjuster clicks may

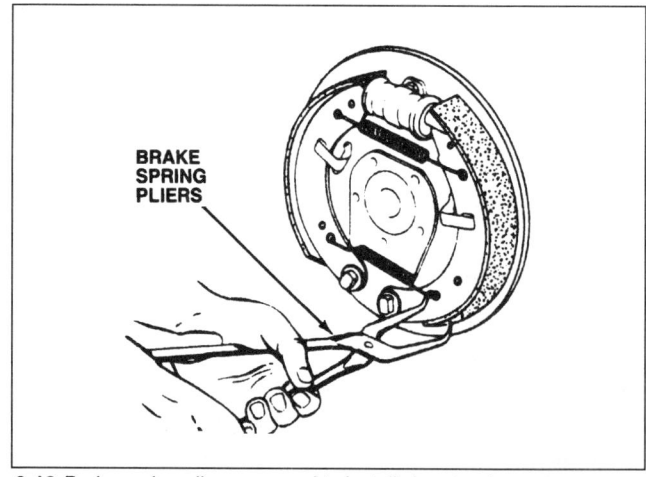

2-19 *Brake spring pliers are used to install shoe-to-shoe return springs.*

be required to get the proper clearance. This is both tiring and time-consuming. When you make the initial adjustment before you install the drum, final manual adjustment will be quick and easy, or you can use the automatic adjusters to make the final adjustment during the test drive.

To perform the initial adjustment, place a shoe-setting caliper inside the brake drum and slide the tool back and forth as you spread the jaws until they span the drum at its widest point, figure 2-20. Then, tighten the lock screw to fix the caliper at this setting. Depending on the brand of tool being used, the opening on the opposite side of the caliper is now set to either equal the drum diameter or at approximately 0.020 inch (0.50 mm) smaller than the drum diameter, pro-

viding a clearance of 0.010 inch (0.25 mm) between the drum and each brake shoe.

Remove the caliper from the drum and place the open side over the brake shoes, 2-21. If the caliper opening matches the drum diameter, rotate the starwheel adjuster as needed until there is approximately 0.020 inch (0.50 mm) clearance between the caliper opening and the shoes at their widest point. If the caliper setting includes the desired lining-to-drum clearance, rotate the starwheel adjuster as needed until the caliper just slides over the shoes at their widest point. Hold the automatic adjuster pawl out of the way while turning the starwheel so it does not become burred. On brakes with dual starwheel adjusters, rotate each starwheel an equal amount.

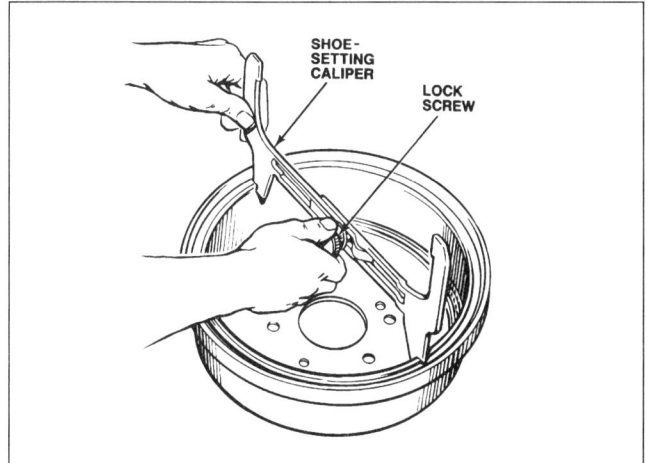

2-20 *Fit the shoe-setting caliper inside the brake drum and adjust it to fit the widest, inside diameter of the drum.*

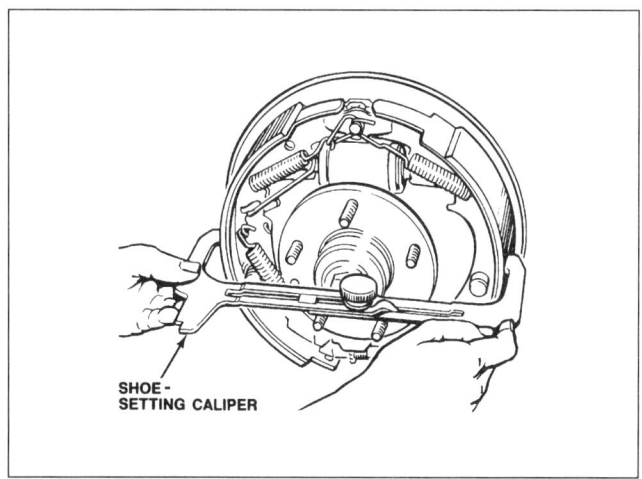

2-21 *Transfer the drum diameter to the shoes by fitting the opposite jaws of the caliper over the brake friction assembly.*

1. During a road test, a loud, scraping noise occurs as the brakes are released. This is most likely caused by:
   a. Worn out brake linings
   b. A cracked brake drum
   c. Defective brake shoes return springs
   d. A bent backing plate

2. Which of the following is **NOT** used to hold floating drums in place?
   a. Speed nuts
   b. Bolts threaded into the hub
   c. The wheel and lug nuts
   d. A large nut that preloads the wheel bearing

3. Fixed drums are held in place by:
   a. Speed nuts
   b. Bolts threaded into the hub
   c. A large nut that preloads the wheel bearing
   d. The wheel and lug nuts

4. Technician A says that brake drums should always be replaced if they are cracked. Technician B says that brake drums should be reinstalled on the wheel opposite from where they were removed. Who is right?
   a. A only
   b. B only
   c. Neither A nor B
   d. Both A and B

5. Visually inspect brake drums for:
   a. Thickness variation
   b. High spots
   c. Heat checking
   d. Out-of-round

6. Which of the following CANNOT be measured with an inside micrometer?
   a. Drum inside diameter
   b. Drum taper, barrel, and bellmouth distortion
   c. Out-of-round distortion
   d. Eccentric drum distortion

7. After turning, a brake drum must measure how much smaller than its discard diameter if it is reusable?
   a. 0.300 inch (7.5 mm)
   b. 0.030 inch (0.75 mm)
   c. 0.003 inch (0.075 mm)
   d. 0.0003 inch (0.0075 mm)

8. An out-of-round brake drum can cause all of the following **EXCEPT**:
   a. Pedal pulsation
   b. Brake vibration
   c. Erratic braking
   d. Excessive pedal travel

9. After turning, the inside diameters of brake drums on the same axle should measure within:
   a. 0.010 inch (0.25 mm) of each other
   b. 0.020 inch (0.50 mm) of each other
   c. 0.030 inch (0.75 mm) of each other
   d. 0.040 inch (1.00 mm) of each other

10. When machining both drums from an axle, which drum should be turned first?
    a. The most worn
    b. The least worn
    c. Either one
    d. The first one removed

11. When disassembling a drum brake, after the drum is removed, the next step is generally to remove the:
    a. Shoes
    b. Starwheels
    c. Return springs
    d. Wheel cylinders

12. When performing a complete drum brake job, the brake hardware should be:
    a. Reused
    b. Replaced
    c. Cleaned and inspected
    d. Measured

13. When assembling a drum brake, apply high-temperature brake grease to the:
    a. Contact pads
    b. Shoe webs
    c. Parking brake cables
    d. Return spring anchors

14. Inspect a wheel cylinder for all of the following **EXCEPT**:
    a. Free movement of the pistons
    b. Rigid mounting
    c. Leakage
    d. Hard spots and heat checks

15. Set the initial brake adjustment with a:
    a. Shoe-setting caliper
    b. Brake micrometer
    c. Screw driver after the drum is installed
    d. Torque wrench

Chapter Three

DISC BRAKE SERVICE

There are three main categories of disc brake service: pad replacement, caliper overhaul, and rotor machining or replacement. Although it is impossible to determine what repairs are needed until the brake assemblies are inspected, a preliminary road test can help pinpoint problems.

During the road test, listen for noises and note any pedal pulsation or vehicle pulling when applying the brakes. Noises indicate brake pad problems, pedal pulsation results from rotor damage, and pulling is due to uneven brake application, which is caused by a defective caliper.

DISC BRAKE NOISES

Disc brakes can generate a number of noises as the brakes apply, most are the result of friction material problems. A scraping or grinding sound indicates worn out brake linings that have resulted in metal-to-metal contact between the pad backing plate and rotor. Squeaks and squeals occur when the brakes are applied and often result from either problems with the lining friction material, or worn, damaged, and missing vibration damping or anti-rattle parts. A snap or click on brake application is caused by loose, worn, or broken mechanical parts, as well as an improper rotor or drum surface finish. The exact cause is determined during the brake system inspection.

On some disc brakes, a light scraping or chirping noise when the brakes are not applied is caused by the pad wear indicator rubbing against the rotor, signaling the need for pad replacement, figure 3-1. A louder, scraping noise can be caused by a bent splash shield rubbing a rotor.

BRAKE PAD REPLACEMENT

Brake pads are sold and serviced as axle sets, which consist of four pads, the inner and outer pad for the caliper at each wheel. Never mix pads from different manufacturers, as the friction coefficients of the linings may be different, even though they appear the same.

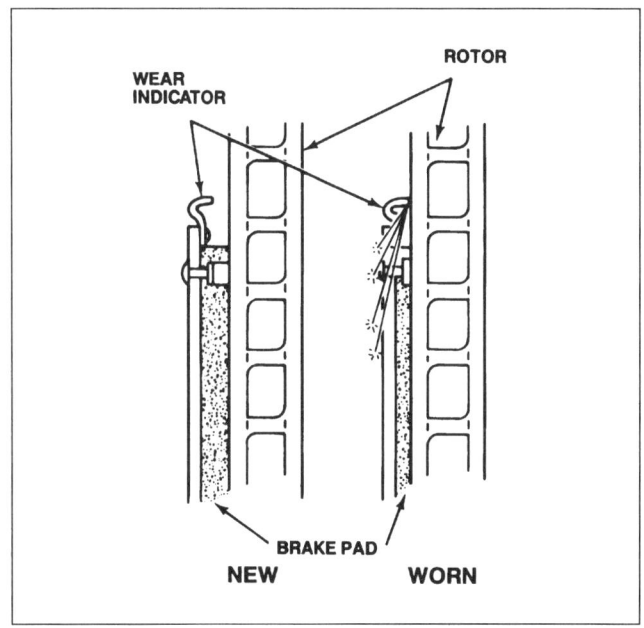

3-1 *Brake wear indicators make a constant chirping noise to notify the driver when the friction material is worn and the pads need replacement.*

If only one pad of an axle set is badly worn, replace the entire set after repairing the problem that caused the uneven wear.

Most manufacturers recommend that the bushings, O-rings, retaining bolts, retaining clips, and any other caliper mounting hardware be replaced whenever a new set of brake pads is installed. The exact procedure for replacing a set of brake pads varies with the design of the caliper. However, there are a number of basic steps common to any pad replacement.

Pad Removal

The first step in replacing brake pads is to remove some of the brake fluid from the master cylinder reservoir with a brake fluid syringe. This makes space for the fluid that will be displaced back into the reservoir, when the caliper pistons are bottomed in their bores, to install the new, thicker pads.

On **fixed calipers**, remove the pad guide pins and retaining spring, then use a pair of pliers to pull the pads straight out of the caliper, figure 3-2. If there is a ridge at the edge of the rotor that prevents a pad from being easily removed, insert a screwdriver between the rotor and pad, and carefully pry the caliper piston back into its bore until there is sufficient clearance to remove the pad.

Removing the pads from **sliding calipers** or **floating calipers** generally requires that the movable portion of the caliper be separated or pivoted away from the anchor plate, figure 3-3. Whenever you remove a caliper during brake pad replacement, hang it from the suspension by a wire, so there is no strain on the brake hose that might cause internal or external damage.

Brake Pad Inspection

The thickness of the lining material is the main factor that determines whether the pads should be replaced. The brake pads must be at least 0.030 inch (0.75 mm) above the pad backing plate or rivet heads, figure 3-4. Measure friction thickness with a machinist scale.

Also, inspect the pads for taper wear, in which the pads are thinner at one end than at the other. Some pad taper wear is normal in floating calipers because the caliper body tends to flex slightly on its mountings. The leading edges of brake pads may also wear faster than the trailing edges because they operate at higher temperatures. However, if there is more than 0.13 inch (3 mm) of taper wear, you should replace the pads and inspect the caliper for possible problems.

Compare the amount of wear on the two pads in each caliper, then compare the amount of pad wear between the two calipers on the same axle. Uneven wear between pads in the same caliper can be caused if the rotor is rough on one side, causing that pad to wear more rapidly. In fixed calipers, a frozen piston will cause uneven wear between the two pads.

All of the pads may not be worn the same amount in sliding or floating calipers because the pad on the piston side usually wears more quickly. However, grossly uneven pad wear occurs in floating and sliding calipers when the mounting hardware rusts or corrodes, causing the caliper to bind as it moves on the guide pins or anchor plate.

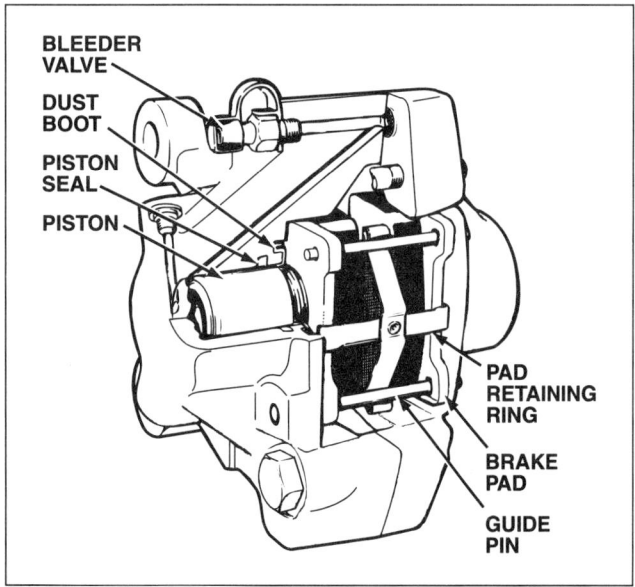

3-2 *A fixed caliper rigidly mounts to the chassis. Once the guide pins and retaining spring are removed, the pads can be pulled free of the caliper.*

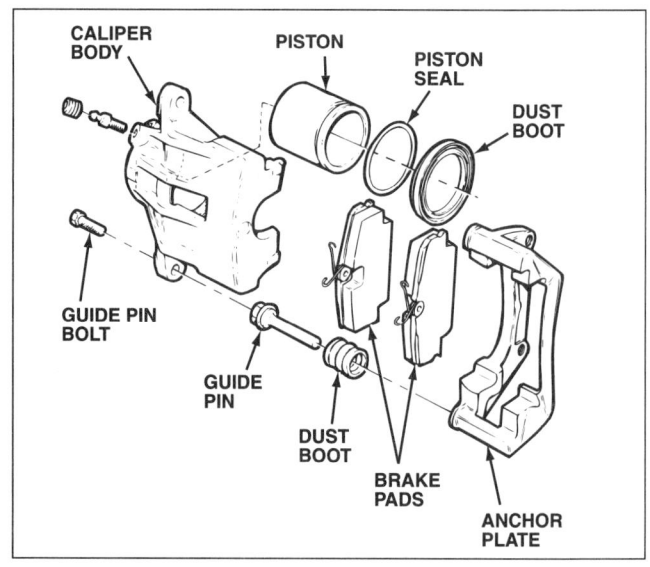

3-3 *With floating calipers, the anchor plate attaches to the chassis, and the caliper body slides on guide pins connected to the anchor plate as the brakes are applied. To replace the pads, the caliper body must be separated from the anchor plate.*

Fixed Caliper: A brake caliper that solidly bolts to the vehicle suspension. The caliper does not move when the brakes are applied.

Sliding Caliper: A two-piece brake caliper consisting of a body and anchor plate. The anchor plate rigidly attaches to the vehicle suspension, and the body slides on machined ways to bring the pads into contact with the rotor.

Floating Caliper: A two-piece brake caliper consisting of a rigid anchor plate and movable body that compresses the pads as the brakes apply. The caliper body is supported by bushings and O-rings that slide on guide pins and sleeves, which attach to the caliper anchor plate.

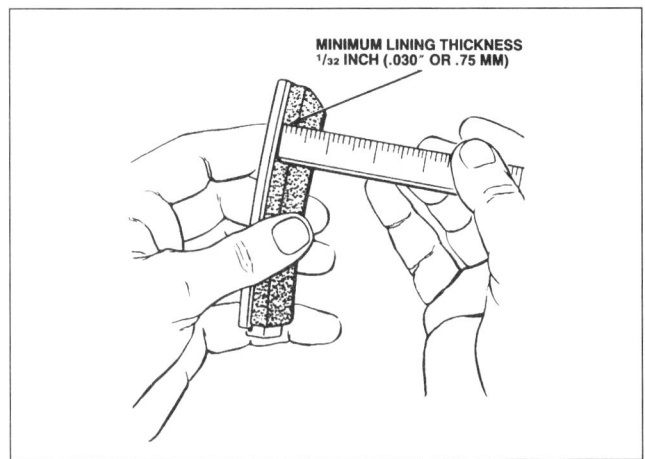

3-4 *Measure the thickness of the friction material on brake pads to determine if they can be returned to service.*

Check the friction lining surface for signs of contamination from brake fluid that has leaked past the piston seal and dust boot. Fluid contamination will cause the lining to darken. Brake pads that have been soaked with brake fluid must be replaced.

Inspect the pads for physical damage. Look for large cracks in the lining, loose or missing rivets, a bent backing plate, or a bonded lining that is separating from the backing plate, figure 3-5. If these problems are present, replace the pads.

Pad Installation

Always inspect the brake rotor and the brake caliper, as described later in this chapter, before installing new pads. The caliper external inspection procedure also describes how to bottom the caliper pistons in their bores, so that the new brake pads will fit.

Install the brake pads into the caliper or anchor plate as dictated by the brake design. The pads in fixed calipers slip into place, and a spring retainer on the guide pins prevents the pads from vibrating and causing brake noise. Some designs use shims that fit between the caliper pistons and pad backing plates to reduce vibration and noise, figure 3-6. The pads in floating and sliding calipers usually have spring clips or bent tabs on the backing plate that lock them securely into the caliper anchor plate. Antirattle spring clips, of which a number of designs are used, attach to the brake pads or caliper anchor plate to reduce pad vibration and prevent brake noise.

Once the pads are in position, secure them in the appropriate manner. With a fixed caliper, install the guide pins and retaining spring. With a sliding caliper, clean the **ways** and lubricate them using a high-temperature brake grease, figure 3-7. Then, position the caliper body onto the ways over the rotor, and install the retaining hardware. With a floating caliper, lightly coat the caliper bushings and mounting bolts or guide pins with high-temperature brake grease, then position the caliper body over the rotor. Install the mounting bolts and tighten all fasteners to specified torque.

BRAKE CALIPER EXTERNAL INSPECTION

Inspect the entire outside of the caliper body. Replace the caliper if you find cracks or other major damage. Visually inspect the caliper piston dust boot. It must be

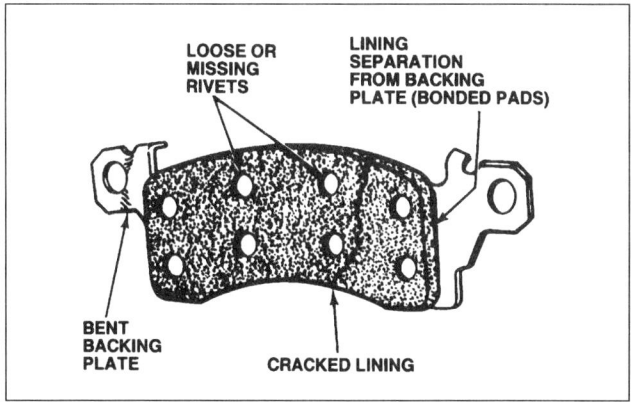

3-5 *Inspect brake pads for physical damage, such as lining cracks, loose or missing rivets, backing plate distortion, or separation from the backing plate, as well as wear and fluid contamination.*

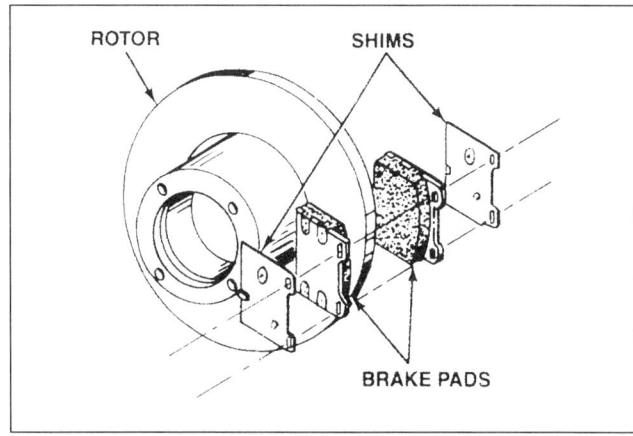

3-6 *Shims install behind the brake pad backing plates on some vehicles to reduce pad vibration and noise.*

Ways: Polished, machined surfaces that permit movement between two metal parts. Ways are machined into the anchor plate and caliper body that provide a sliding surface for the caliper.

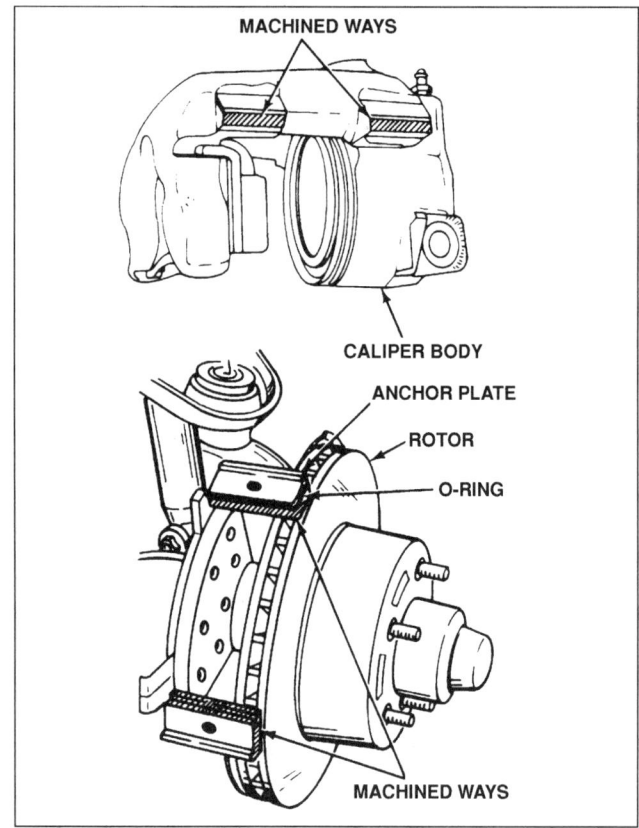

3-7 *Make sure the machined ways of a sliding caliper are clean and smooth, then lubricate them with a thin coat of high-temperature brake grease.*

fully seated in the caliper body and be free of holes and tears. Also, inspect the caliper for brake fluid leakage around the boot and look for fluid deposits on the brake pads that can indicate a leaking caliper seal. If the dust boot is unseated or damaged, or there are signs of fluid leakage, rebuild or replace the caliper.

If the dust boot is in good condition and there are no signs of leaks, bottom the caliper pistons in their bores. A C-clamp can be used to bottom the piston on most single-piston sliding calipers, figure 3-8. A large pair of

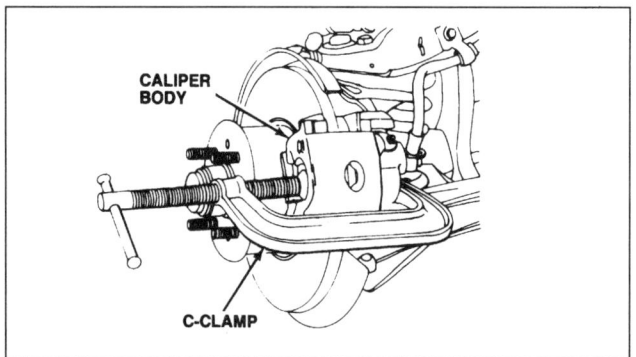

3-8 *Using a C-clamp to bottom the piston of a single-piston sliding caliper.*

slip-joint pliers can be used to seat the piston on some floating and sliding calipers, figure 3-9. Take care not to tear or dislodge the dust boot. On multi-piston fixed calipers, insert a screwdriver or pry bar between the rotor and an old brake pad and lever the pistons back into their bores. Avoid prying against the machined contact surface of the rotor.

Rear disc brake calipers that apply to serve as the parking brake may require special tools to seat the piston. Typically, a turning tool is used to thread the piston back into the bore on the automatic adjuster screw, figure 3-10. On assembly, the pad to rotor clearance may need to be checked and adjusted as well.

When bottoming the pistons, note the feel of the piston movement. A moderate amount of force should

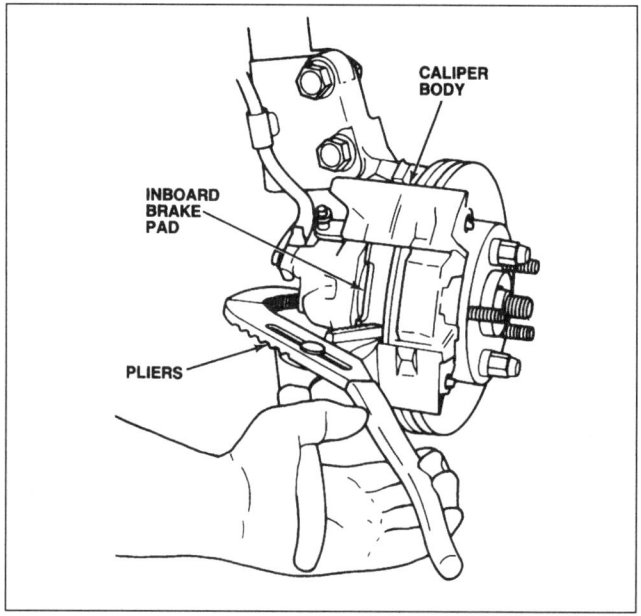

3-9 *Using slip-joint pliers to bottom the piston in a floating caliper.*

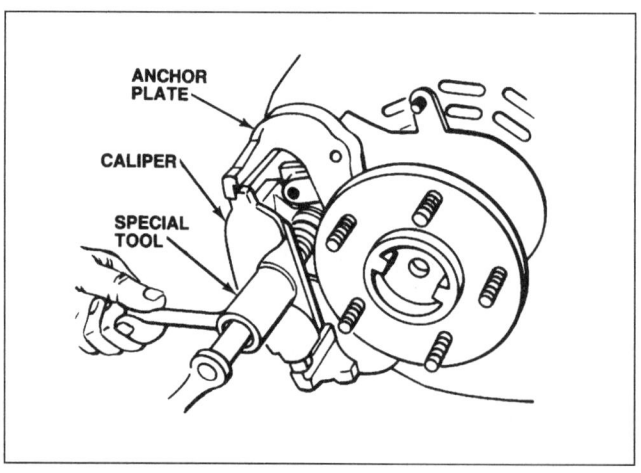

3-10 *Special tools may be required to retract the piston of a rear brake caliper that contains the parking brake mechanism.*

be sufficient to move the pistons, and they should slide smoothly into the caliper bores. Furthermore, all of the pistons in a multi-piston caliper, as well as those in the calipers on both sides of the vehicle, should require about the same amount of force to bottom. If a piston is frozen or difficult to bottom, there is rust or corrosion on the piston or in the bore, and the caliper must be overhauled or replaced. Just as brake pads are always replaced in axle sets, both calipers on an axle should be overhauled or replaced if it is determined that one needs to be serviced.

BRAKE CALIPER OVERHAUL

Every brake caliper requires certain special overhaul procedures, but just as with brake pad replacement, all overhaul procedures have some things in common. The following sections describe a general procedure. Before beginning the overhaul, attempt to loosen the caliper bleeder valve. If the bleeder valve cannot be loosened, replace the caliper.

Caliper Piston Removal

Special tools are available to grip and mechanically remove brake caliper pistons. However, it is more common to use compressed air or hydraulics to free pistons from their bores.

Compressed-air piston removal

Sufficient force is available from the compressed air supply in most shops to remove all but severely frozen pistons. Use extreme caution when removing caliper pistons in this manner. Clamp the caliper in a vise by its mounting flange. Insert a wooden block or bundle of shop towels between the piston and caliper body to prevent damage to the piston when it comes out of the caliper bore. Remove the bleeder valve and place a

rubber-tipped air nozzle into the caliper fluid inlet. Then, slowly apply air pressure to force the piston out of the bore, figure 3-11. Use the minimum amount of pressure necessary.

Hydraulic piston removal

Using hydraulic pressure to remove caliper pistons is both the most effective and the safest method.

To remove a caliper piston using the brake hydraulic system with fixed brake calipers, remove the brake pads from the calipers. If the brake rotor is not thick enough to prevent the pistons from coming out of their bores, insert a set of worn-out brake pads or thin, wooden shims between the caliper pistons and the rotor.

With floating or sliding calipers, remove the calipers from over the brake rotors, and hang them from the suspension to prevent damage to the brake hose. Insert thick wooden blocks into the calipers between the pistons and caliper bodies to prevent the pistons from coming completely out of their bores.

Place drain pans under the calipers to catch brake fluid in the event a piston does come out of its bore. Slowly apply the brake pedal to force the pistons from their bores.

Dust Boot and Piston Seal Removal

Once the piston is out of the caliper, remove the dust boot. If the boot is a press fit in the caliper and remains attached to it, use a screwdriver to gently pry the boot out, figure 3-12. Take care not to scratch the caliper bore as you remove the boot. If the boot is a stretch fit over the piston and remains attached to that part, simply pull it free and set it aside.

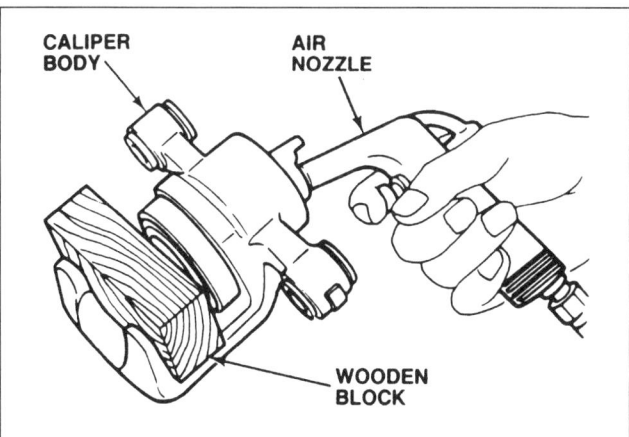

3-11 *A wooden block helps prevent damage when compressed air is used to remove a caliper piston from its bore.*

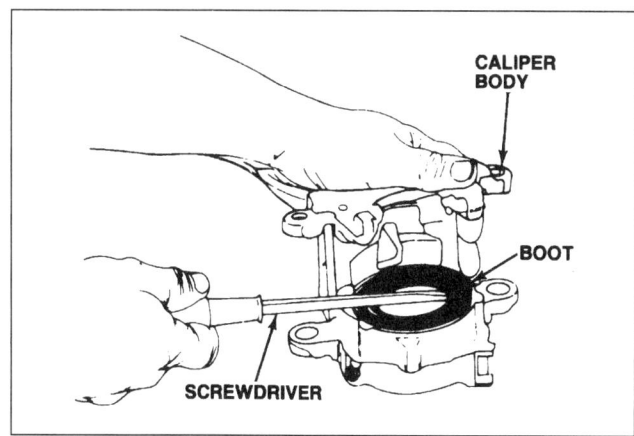

3-12 *Take care to avoid scratching the bore when using a screwdriver to remove a piston dust boot that is press fit into the caliper body.*

Next, remove the piston seals. If the caliper has **stroking seals**, insert a probe under the seals to lever them off of the pistons. If the caliper has **fixed seals**, insert a probe under the seals to pry them out of the grooves in the caliper bores, figure 3-13. To avoid damaging the pistons, caliper bore, or seal groove, always use a wooden or plastic probe to remove seals. Damage in these areas can cause fluid leaks when the caliper is reassembled.

Once the caliper is disassembled, clean the body and pistons with clean brake fluid. If the parts are particularly dirty, use a soft, bristle brush to help remove contamination. Do not use a wire brush.

BRAKE CALIPER INTERNAL INSPECTION

Inspect the caliper bore for scoring, rust, corrosion, and wear. This is particularly important on calipers that have stroking seals. Since the bore provides the sealing surface for a stroking seal, even a small imperfection in the bore can create a fluid leak. Minor rust and corrosion can be removed by polishing with crocus cloth or by honing, which is described later in the chapter. If the bore of a caliper with stroking seals is significantly damaged, the caliper must be replaced.

The bore condition of calipers with fixed seals is less critical because the piston provides the sealing surface. Inspect the bores of these calipers for major defects that might prevent the piston from moving freely, and make sure the edges of the seal groove are not rusted or corroded and do not have nicks or burrs that can cut the seal or affect its sealing. As long as

the seal groove is in good condition and the bore diameter remains within specifications, calipers with fixed seals can be honed clean of even major rust and corrosion. However, if the seal groove is damaged or honing makes the bore oversize, the caliper must be replaced.

Caliper Piston Inspection

Caliper pistons are made of cast-iron, steel, aluminum, or phenolic resin. As mentioned above, pay particular attention when you inspect pistons used in calipers with fixed seals because the outside diameter of the piston provides the sealing surface for the caliper seal. On pistons used with stroking seals, the important area is the groove in which the seal fits. In both cases, the piston must be replaced if the critical surface is not smooth, clean, and entirely free of defects.

Inspect cast-iron, steel, and aluminum pistons for rust, corrosion, nicks, and scoring in their sealing areas. If the piston is chrome plated, make sure the plating is not flaking away. Aluminum pistons are anodized to provide a durable finish, and this surface should not be damaged in any way. If cast-iron, steel, or aluminum pistons have any of these problems, you should replace them.

Pistons made of phenolic resin are prone to cracking and chipping and must be thoroughly inspected, figure 3-14. Minor cracks or chips that extend partially across the piston face are acceptable, as are minor

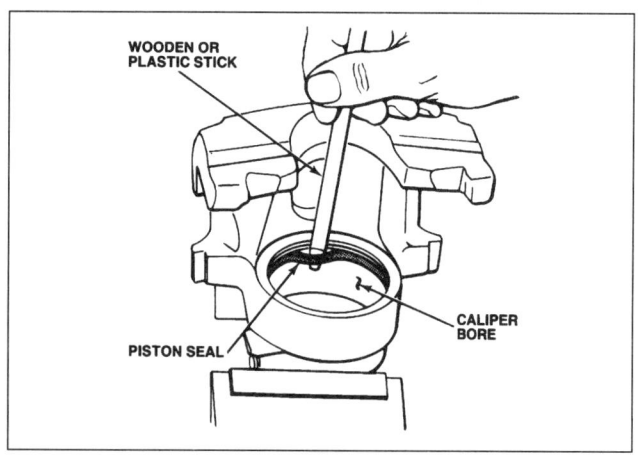

3-13 *Avoid damaging the caliper bore or piston by using a plastic or wooden probe to remove piston seals.*

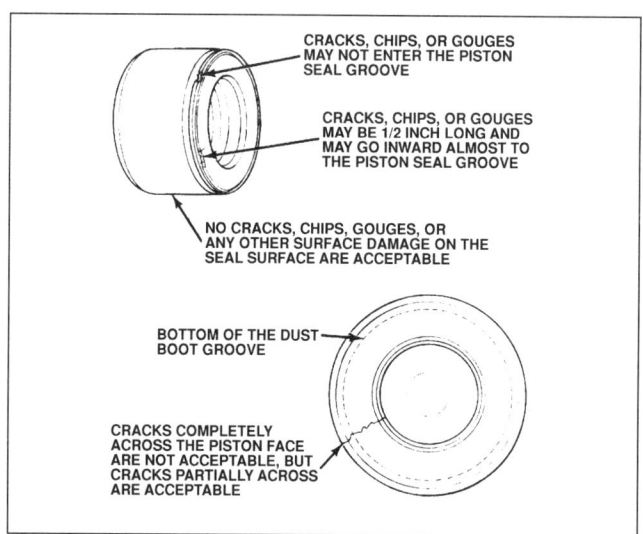

3-14 *Carefully inspect phenolic caliper pistons for cracks, chips, and gouges.*

Stroking Seal: A rubber seal that fits into a groove on the caliper piston and strokes, or moves, with the piston as the brakes apply and release.

Fixed Seal: A rubber seal that fits in a groove machined into the caliper. The caliper piston slides through the inside of the seal as the brakes apply and release.

nicks and gouges at the outer edge of the piston, providing they do not extend into the dust boot groove. Replace a piston if cracks extend across the face of the piston or nicks and gouges enter the dust boot groove. Finally, inspect the outer diameter of the piston. It should be smooth and even. If the surface is scuffed or scored, replace the piston.

BRAKE CALIPER HONING

To hone a brake caliper, follow these steps:

1. Mount the caliper in a vise so you have clear access to the bore. Do not clamp on the cylinder body, this can distort the bore.
2. Select the proper size caliper hone and chuck it in a drill motor. Use a hone with fine stones.
3. Lubricate both the bore and hone with clean brake fluid, place the hone into the bore, and operate the drill motor at approximately 500 rpm for about 10 seconds, figure 3-15. If the caliper uses stroking seals, stroke the hone gently in and out of the bore to obtain a crosshatch pattern for good sealing. If the caliper uses fixed seals, you can simply hold the hone steady in the bore because the honing pattern does not matter.
4. Remove the hone, wipe the bore clean, and check the finish. It should be clean, smooth, and free of damage. Repeat the honing sequence as needed. If the bore does not clean up after several attempts, replace the caliper.
5. After you are finished honing, thoroughly clean the caliper bore, seal groove, and fluid inlet passage with brake cleaner. Allow the caliper to air dry, and make sure all traces of grit and residue are cleaned away.

Caliper Bore Measurement

There is usually little or no caliper piston and bore wear during normal operation. However, after honing a caliper, the bore must be measured to make sure it has not been honed too far oversize. This step is critical when honing calipers with fixed seals or when a great deal of honing is required to clean up the bore.

Most manufacturers specify between 0.004 and 0.010-inch (0.10 and 0.25 mm) piston-to-bore clearance for calipers with stroking seals. Consult the factory shop manual for the vehicle you are servicing to get an exact figure.

Too much clearance in a caliper with fixed seals results in seal damage, and the self-adjusting action of the caliper may be affected as well. As a general rule, the maximum piston to bore clearance for fixed seal calipers is 0.002 to 0.005 inch (0.06 to 0.13 mm) for metal pistons and 0.005 to 0.010 inch (0.13 to 0.25 mm) for phenolic pistons. Consult the factory shop manual for an exact figure.

To measure the caliper piston-to-bore clearance, insert a feeler gauge blade that equals the maximum allowed clearance into the caliper bore, figure 3-16. Then, attempt to insert the piston into the bore alongside the feeler gauge blade. If the piston can enter the bore, repeat the check with a new piston. If a new piston can also enter the bore, the bore is oversized, and the caliper must be replaced.

BRAKE CALIPER ASSEMBLY

After disassembling, inspecting, cleaning, honing, and measuring the caliper and caliper piston, obtain the replacement parts needed along with the correct overhaul and small parts kits.

Piston Seal Installation

On calipers with fixed seals, thoroughly lubricate the O-ring seal and the seal groove in the caliper bore with fresh brake fluid. Position one edge of the seal into the

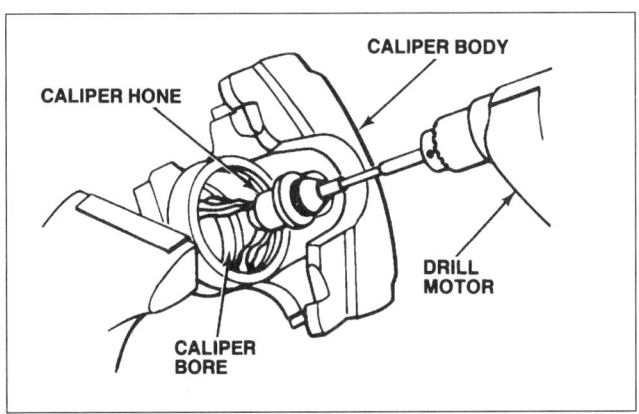

3-15 *Honing restores the surface of the caliper piston bore.*

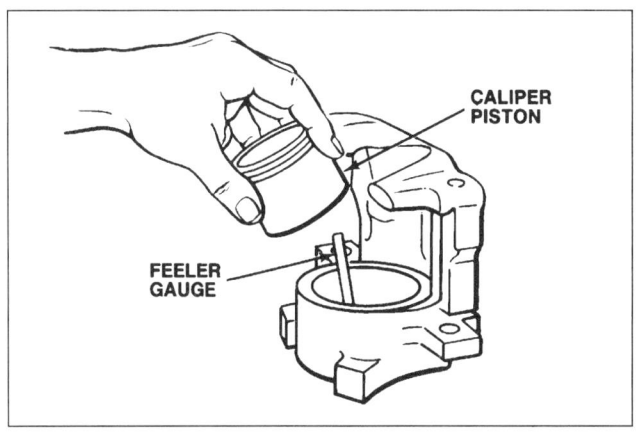

3-16 *Checking piston-to-bore clearance with a feeler gauge. The feeler gauge equals the maximum allowable clearance.*

groove, then gently work the seal around the bore diameter with your fingers until it is fully seated, figure 3-17. Do not twist or roll the seal. Square-cut O-ring seals can be installed facing either way. However, if the seal is a shape other than square-cut, install it facing in the same direction as the original part.

On calipers with stroking seals, thoroughly lubricate the lip seal and the seal groove in the caliper piston with fresh brake fluid. Carefully stretch the seal over the end of the piston and into place in the seal groove by hand, figure 3-18. Take special care not to cut or damage the seal, particularly the edge of the sealing lip. The sealing lip of a stroking seal must face the bottom of the caliper bore when the piston is installed.

Press-fit Dust Boot Installation

Some calipers use a dust boot with a metal reinforcing ring around the outer edge that is a press fit into the caliper body. These boots require a special seal driver to properly install them. To assemble:

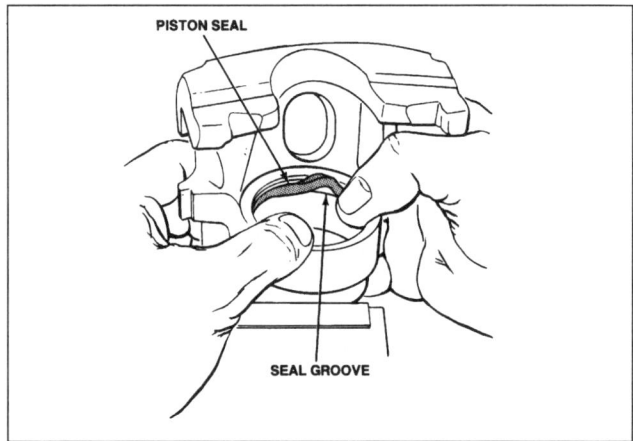

3-17 *Use fresh brake fluid to lubricate a fixed seal, then install it into the caliper bore groove by hand.*

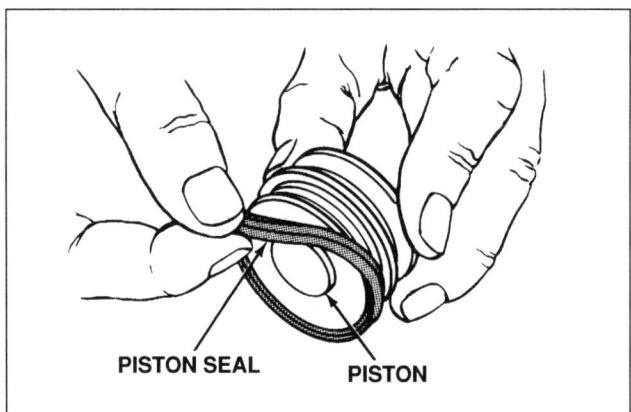

3-18 *Avoid overstretching a stroking seal as you work it over the piston and into its groove by hand.*

1. Lubricate the piston with fresh brake fluid, then slip it into the dust boot until the boot snaps into the groove of the piston.
2. Using only hand pressure, work the piston into the caliper bore past the O-ring seal, and bottom the piston in the bore.
3. Select a dust boot driver of the proper size and use a hammer to seat the metal reinforcing ring at the outer edge of the boot into the groove in the caliper body, figure 3-19.

Retaining Ring Dust Boot Installation

Many imports use a dust boot with a separate metal ring that secures the outer edge of the boot into the caliper body. No special tools are required to assemble this type of caliper. To assemble the piston and dust boot on these calipers:

1. Lubricate the piston with fresh brake fluid and slip it into the dust boot until the boot snaps into the groove in the piston.
2. Work the piston into the caliper bore past the O-ring seal by hand until the piston bottoms in the bore.
3. Using your fingers, position the outer edge of the boot into the groove in the caliper body; the boot must seat properly.
4. Install the metal ring to lock the boot into the groove, figure 3-20.

Lip/Groove Dust Boot Installation

Some brake designs use a dust boot with a lip around the outer edge that fits into a groove in the caliper

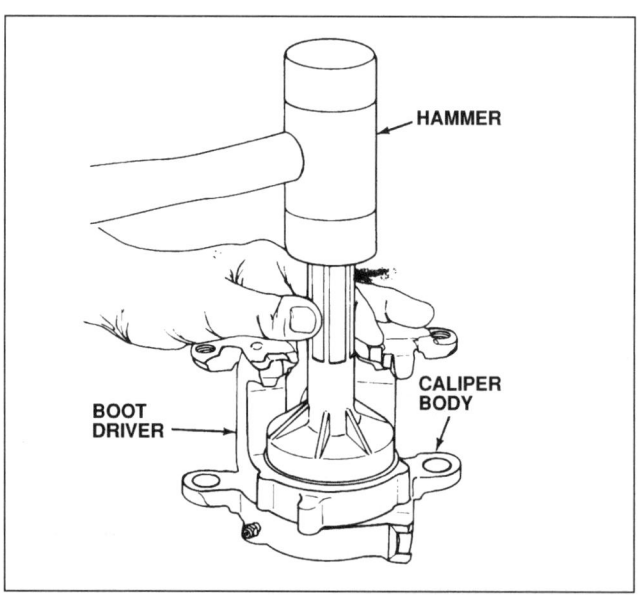

3-19 *A boot driver is required to seat a dust boot with a metal reinforcing ring into the caliper body.*

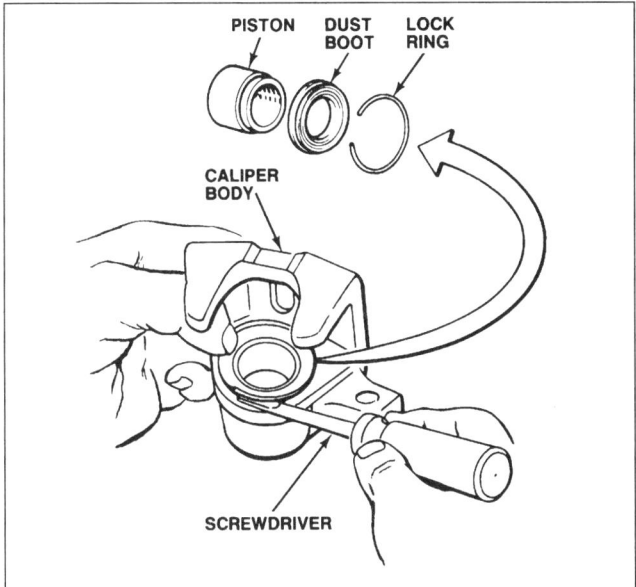

3-20 *On a caliper boot with a metal lock ring, fit the piston and dust boot by hand, then work the ring over the boot so it snaps into place.*

bore. Although special tools are not essential to assemble these calipers, a set of metal or plastic rings about a $1/2$-inch high will make the job easier. These rings are readily available from tool supply outlets. To assemble the pistons and dust boots on these calipers:

1. Select the ring that just barely fits over the caliper piston.
2. Lubricate the inner edge of the dust boot with fresh brake fluid, then install it over the assembly ring.
3. Install the lip at the outer edge of the dust boot into the groove in the caliper bore. Reach through the ring with your fingers to make sure the lip is fully seated in the groove.
4. Lubricate the piston with fresh brake fluid, then slip it through the ring into the caliper bore. Work the piston past the O-ring seal by hand until it is bottomed in the bore.
5. Carefully remove the ring from the dust boot so that the boot snaps into place on the piston.

BRAKE ROTOR INSPECTION

Brake rotor inspection consists of two parts, a visual inspection followed by one or more careful measurements. A thorough inspection determines if the rotor is in serviceable condition, must be machined to restore its surface, or is beyond saving and must be replaced.

BRAKE ROTOR VISUAL INSPECTION

To begin, wipe the rotor surface clean to make it easier to spot problems. Like the drum visual inspection

detailed previously, look the rotor over for problems on the surface, such as:
- Scoring and grooves
- Cracks
- Heat checking
- Hard spots.

Rotors must be resurfaced if scoring and grooves are deeper than 0.010 inch (0.25 mm). If cracks are visible, replace the rotor. Minor, localized heat checks can be machined out, but if heat checking is widespread, replace the rotor. As with drums, hard spots are difficult to remove, and most manufacturers recommend replacing rotors that have hard spots.

Brake Rotor Measurement

Brake rotors are measured to identify wear and distortion that is not visually apparent. All types of rotor wear and distortion can be measured using an outside micrometer and a dial indicator.

Rotor thickness

The first step is to measure rotor thickness with an outside micrometer to check for wear. To begin, note the discard dimension stamped or cast into the rotor. Then, take a micrometer reading 1 inch (25 mm) in from the outer edge of the rotor to measure thickness, figure 3-21. For best results, use a brake micrometer with a pointed anvil that will measure to the bottom of wear grooves. Compare the measured thickness to the discard dimension.

If the rotor thickness is not at least 0.015 to 0.030 inch (0.40 to 0.75 mm) larger than the rotor discard dimension, replace the rotor. The amount of additional metal required to allow for wear in service varies depending on the manufacturer. Check the shop manual of the vehicle you are servicing for the exact value. If the rotor needs to be turned, there must be sufficient metal remaining, so the thickness will be at least 0.015 to 0.030 inch (0.40 to 0.75 mm) larger than the discard

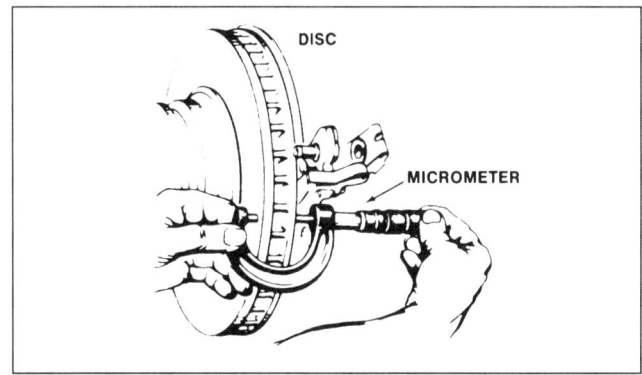

3-21 *Use an outside micrometer and take a reading approximately 1 inch (25 mm) in from the outer edge to measure rotor thickness.*

dimension after machining. Ford Motor Company rotors, and those from some other manufacturers, are marked differently. Replace one of these rotors whenever its thickness is less than the minimum thickness stamped or cast into the outside of the rotor, figure 3-22. If you are unsure of what the measurements on a rotor mean, consult a factory shop manual.

Rotor taper variation

Taper variation is a difference in thickness across the friction surface of a rotor, figure 3-23. To check for taper variation, use an outside micrometer with a deep frame to measure the rotor thickness at the outer edge just below the ridge. Take a second measurement at the inner edge of the area swept by the brake pads. Subtract the smaller measurement from the larger one to obtain the taper variation. Repeat these measurements at four points around the rotor. If the variation is greater than 0.003 inch (0.08 mm) at any point, machine the rotor. A rotor with too much taper will not allow the pads to contact the rotor squarely and can cause the caliper pistons to bind in their bores.

Rotor lateral runout

Lateral runout is a side-to-side movement of the rotor as it turns, figure 3-24. Excessive runout can cause brake pedal pulsations, vibration during braking, and increased brake pedal travel from too much pad knockback. Runout is measured using a dial indicator.

For maximum braking performance, lateral runout should be less than 0.003 inch (0.08 mm). However, lateral runout tolerances vary by manufacturer and, depending on the vehicle, anywhere from 0.002 to 0.008 inch (0.05 to 0.20 mm) may be acceptable. It is only necessary to check lateral runout on one side of the rotor; runout never varies significantly between the two sides.

Check for lateral runout while the rotor is mounted on the vehicle and rotating on the wheel bearings. When making this check, it is very important not to mistake bearing play for lateral runout. Adjustable wheel bearings can be tightened to eliminate play as a factor. With non-adjustable wheel bearings, measure and record bearing play, then subtract it from the final reading on the dial indicator to determine the true runout.

To check the lateral runout of a rotor:

1. Raise and properly support the vehicle so the wheel with the rotor to be checked can turn freely, then remove the wheel.
2. If the vehicle has **floating rotors**, install two lug nuts to hold the rotor tightly on the hub.

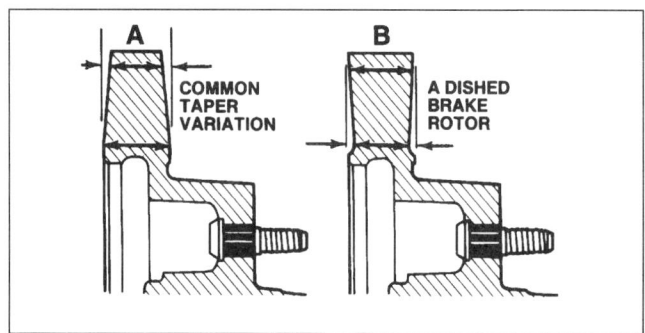

3-23 *Taper variation is uneven wear across the friction surface of a rotor.*

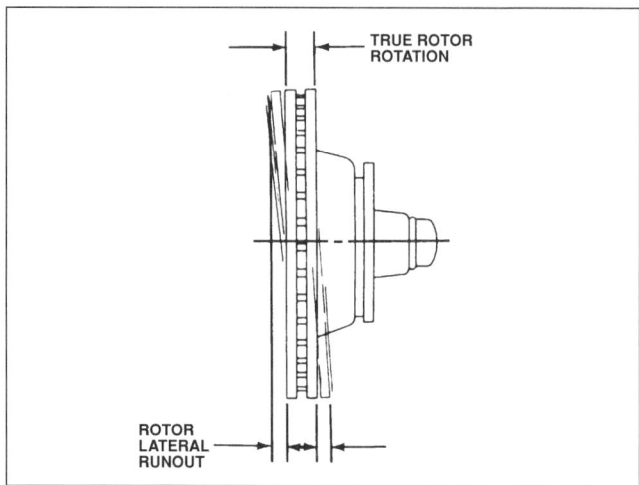

3-24 *Lateral runout causes a rotor to move from side to side, or wobble, as it rotates.*

3-22 *Rotors must be replaced if they measure less than the specified minimum thickness after turning.*

Floating Rotor: A floating rotor is a separate part that installs onto an axle flange or hub assembly. Floating rotors are generally held in place by the wheel lub nuts or bolts. The rotor does not contain the wheel bearings, which are part of the hub.

3. Either pry the brake pads back, so they do not drag against the rotor, or remove the caliper to access the rotor.

4. Mount a dial indicator so the plunger contacts the rotor at a 90-degree angle about 1 inch (25 mm) from the outer edge, figure 3-25.

5. Rotate the rotor until the lowest reading shows on the indicator dial, then zero the dial.

6. Rotate the rotor until the highest reading shows on the dial; this is the lateral runout with adjustable wheel bearings. With non-adjustable wheel bearings, subtract the bearing play from this figure to find the true lateral runout.

Compare findings to specifications to determine if the rotor can be salvaged or must be replaced. A rotor with runout can be machined and returned to service unless turning reduces it to less than minimum thickness.

Rotor lack of parallelism

A rotor that lacks parallelism varies in thickness at different places around its surface, figure 3-26. Often called warpage, lack of parallelism is the most common cause of brake pedal pulsation and also causes braking vibration.

To check for variations in parallelism, use an outside micrometer to measure the rotor thickness at 6 to 12 equally spaced points around the surface. Make all of the measurements at the same distance in from the outer edge of the rotor, so taper variation will not affect the measurements. If the thickness variation between any two points is greater than 0.0005 inch (0.013 mm), and there is noticeable brake pedal pulsation, machine the rotor.

Ford inner surface measurement

On Ford vehicles, an additional measurement must be taken to determine if a rotor can be machined and returned to service. Remove the rotor from the vehicle,

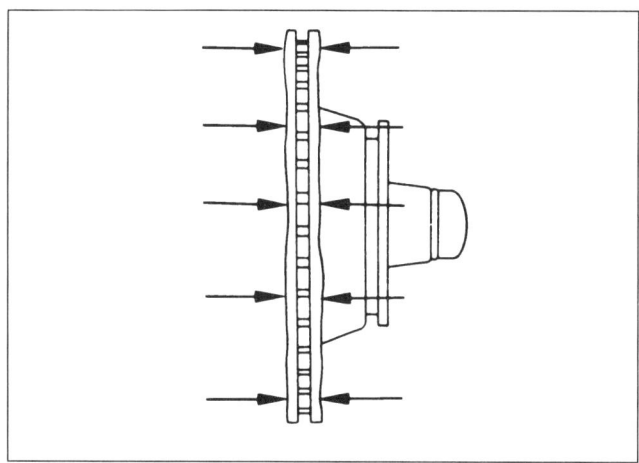

3-26 *Lack of parallelism, or thickness variation, is the most common cause of brake pedal pulsation.*

then measure the inner surface to determine how much, if any, metal can be safely removed. This is done using a depth micrometer and one of two special gauges available from Ford. The gauge for fixed rotors is a large steel ball that fits into the inner bearing race, figure 3-27. For floating rotors, a pyramid-shaped template that fits onto the flange seat is used, figure 3-28.

BRAKE ROTOR MACHINING

When machining rotors, only remove the minimum amount of metal necessary to restore the surface. This helps ensure the longest possible service life for the rotor. On rotors used with sliding or floating calipers, it is acceptable to machine different amounts of metal

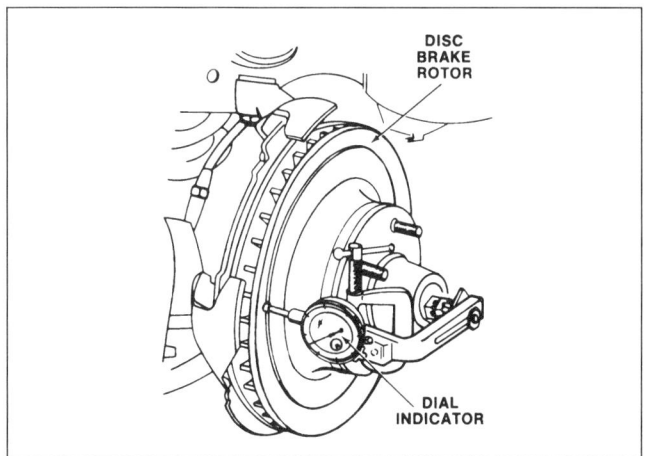

3-25 *Dial indicator setup to measure rotor lateral runout.*

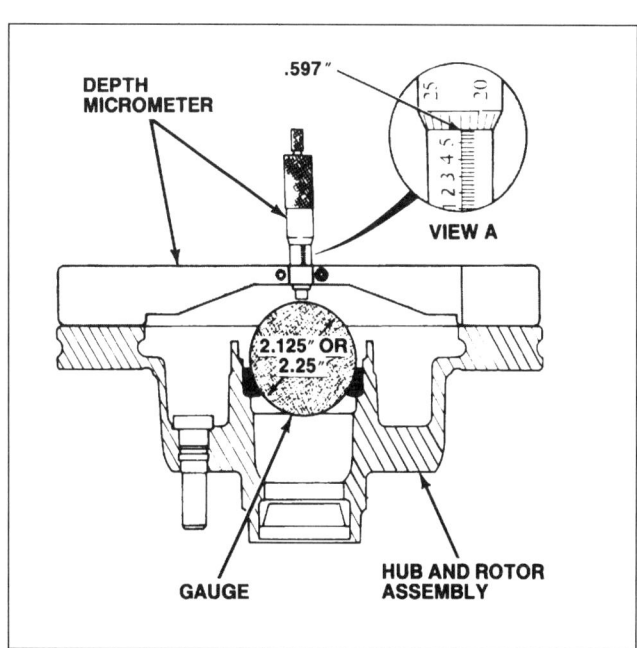

3-27 *A special ball gauge and a depth micrometer are used to measure the inner brake surface on a Ford fixed rotor.*

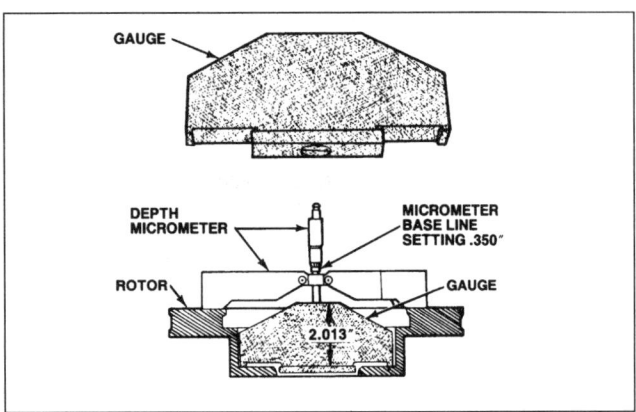

3-28 *Inner brake surface is measured on a Ford floating rotor with a special pyramid-shaped gauge and a depth micrometer.*

from each side. However, rotors used with fixed calipers must be machined equally on both sides.

Never machine a rotor unless you also machine the rotor on the other end of the same axle an equal amount. This keeps braking force and fade resistance equal from side to side and prevents brake pull.

Rotor Resurfacing

Resurfacing is a minor machining operation, in which the surfaces of a brake rotor are ground smooth with a spinning abrasive disc while the rotor is spun on a lathe, figure 3-29. Resurfacing removes very little metal and is not intended to be a substitute for rotor machining. Resurfacing removes rust, brake lining deposits, and minor damage. Resurfacing equipment that restores the friction surface without removing the rotor

from the vehicle is available. This method is the only recommended machining process for some late-model applications.

Resurfacing is also done to put a non-directional finish on rotors after they are machined. When a rotor is turned, the movements of the cutting tools leave almost undetectable spiral grooves across the rotor surfaces. A non-directional finish removes these grooves and helps shorten the break-in period for new pads. This is especially important when fitting semimetallic brake pads that require a non-directional finish to operate efficiently and quietly.

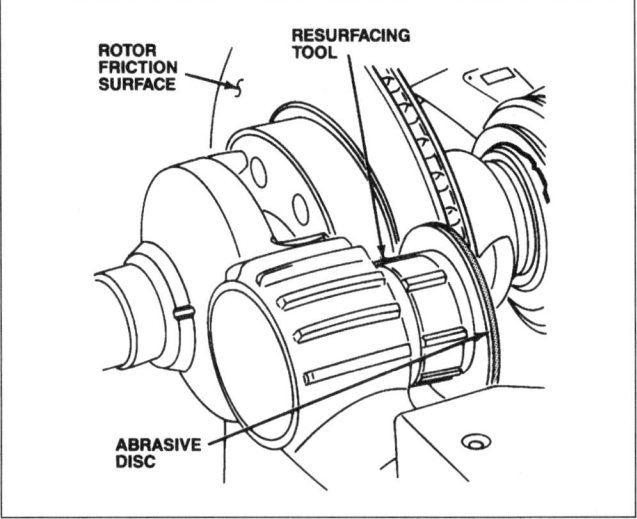

3-29 *Rotor resurfacing uses an abrasive disc to remove rust, brake lining deposits, and minor damage, but not much metal.*

1. An axle set of brake pads consists of:
   a. 8 pads
   b. 4 pads
   c. 6 pads
   d. 2 pads

2. Slight taper wear on disc brake pads removed from a floating caliper is most likely the result of:
   a. A sticking caliper piston
   b. A lack of caliper guide pin lubrication
   c. Normal wear
   d. A lack of rotor parallelism

3. Uneven pad wear in a fixed caliper may be caused by:
   a. A frozen caliper piston
   b. Rust on the caliper ways
   c. Corroded caliper mounting hardware
   d. Normal wear

4. As a general rule, disc brake pads must be replaced if the friction lining thickness measures less than:
   a. 0.020 inch (0.51 mm)
   b. 0.030 inch (0.75 mm)
   c. 0.060 inch (1.59 mm)
   d. 0.130 inch (3.18 mm)

5. To bottom the pistons in a multi-piston caliper, use:
   a. A screwdriver
   b. A C-clamp
   c. Slip-joint pliers
   d. A special tool

6. The safest and most effective way to remove caliper pistons is to use:
   a. Air pressure
   b. A caliper wrench
   c. Hydraulic pressure
   d. A piston press

7. To avoid causing damage, remove caliper piston seals with a:
   a. Brass probe
   b. Small screwdriver
   c. Small steel pick
   d. Plastic or wood probe

8. Caliper bore condition is most critical on calipers with which type of pistons seals?
   a. Fixed
   b. Stroking
   c. Multiple
   d. Square-cut

9. Phenolic caliper pistons should be replaced if an inspection reveals any of the following *EXCEPT*:
   a. Cracks extending across the piston face
   b. Gouges near the dust boot groove
   c. Scuffed or scored outer diameter surface
   d. Minor nicks or gouges at the outer edge

10. For best results, what drill motor speed should be used when honing a caliper piston bore?
    a. 300 rpm
    b. 500 rpm
    c. 1000 rpm
    d. 1200 rpm

11. For calipers with stroking seals, most manufacturers recommend a piston-to-bore clearance of:
    a. 0.004 to 0.010 inch (0.10 to 0.25 mm)
    b. 0.040 to 0.100 inch (1.0 to 2.5 mm)
    c. 0.400 to 0.001 inch (10 to 0.025 mm)
    d. 0.0004 to 0.001 inch (0.010 to 0.025 mm)

12. Which type of caliper requires the least amount of piston-to-bore clearance?
    a. Those with metal pistons and stroking seals
    b. Those with phenolic pistons and stroking seals
    c. Those with metal pistons and fixed seals
    d. Those with phenolic pistons and fixed seals

13. Typically, manufacturers allow rotors to be reused if thickness variation does not exceed:
    a. 0.005 inch (0.127 mm)
    b. 0.002 inch (0.051 mm)
    c. 0.001 inch (0.025 mm)
    d. 0.0005 inch (0.013 mm)

14. Use a dial indicator to check a rotor for:
    a. Lack of parallelism
    b. Taper variation
    c. Lateral runout
    d. Radial runout

15. As a general rule, replace a rotor if, after machining, the measured thickness does not exceed the discard dimension by:
    a. 0.015 to 0.030 inch (0.40 to 0.75 mm)
    b. 0.004 to 0.010 inch (0.10 to 0.25 mm)
    c. 0.0015 to 0.003 inch (0.04 to 0.075 mm)
    d. 0.0004 to 0.001 inch (0.010 to 0.025 mm)

Chapter Four

BRAKE MISCELLANEOUS SERVICE

When diagnosing brake problems or servicing the brakes on a vehicle, it is important to consider all vehicle components that can contribute to brake performance deficiencies. For example, the symptoms of worn or incorrectly adjusted wheel bearings may only be noticeable when the brakes are applied, but servicing the brakes will not cure the problem. A malfunctioning parking brake can cause accelerated or uneven brake friction material wear and other brake performance problems, and while replacing linings cures the symptoms, it does not repair the cause. A thorough diagnosis and inspection is the only way to ensure a complete repair.

WHEEL BEARING INSPECTION

A body vibration that occurs when the brakes are applied during a road test may be caused by loose wheel bearings. However, a worn wheel bearing makes a growling sound when the vehicle is moving, not only when the brakes are applied. To confirm a wheel bearing problem, swerve the vehicle back and forth to alternately load and unload the bearings on opposite sides of the chassis. The noise from the bad bearing will increase with the load on that side of the vehicle and decrease as the load is reduced. Besides bearing noise, there are three other signs that indicate a wheel bearing needs to be serviced: roughness, grease seal leakage, and excessive axial play.

Bearing Roughness

The same wear that causes bearing noise can also be felt as a roughness if the wheel is rotated with the vehicle supported off the ground. If the suspect bearing is not on a driven axle, raise and support the vehicle, then slowly rotate the wheels by hand to check bearing condition. Feel for roughness in the bearing as it turns, and listen for a low-frequency, rumbling noise. Compare the results on opposite sides of the vehicle to determine where the problem lies.

Bearing Grease Seal Leakage

Grease leaking from the bearing grease seal can be a sign that excessive bearing play has caused seal damage. Always replace leaking grease seals to prevent contamination of the brake linings and, at the same time, service the wheel bearings.

Bearing Axial Play

If inward and outward movement of the bearing hub or axle exceeds specifications, the bearings must be serviced. Checking the **axial play**, or endplay, is the only means of determining the condition of sealed, double-row bearing assemblies. The most accurate way to check bearing play is to use a dial indicator. To check:

1. Raise and properly support the vehicle, then remove the wheel.

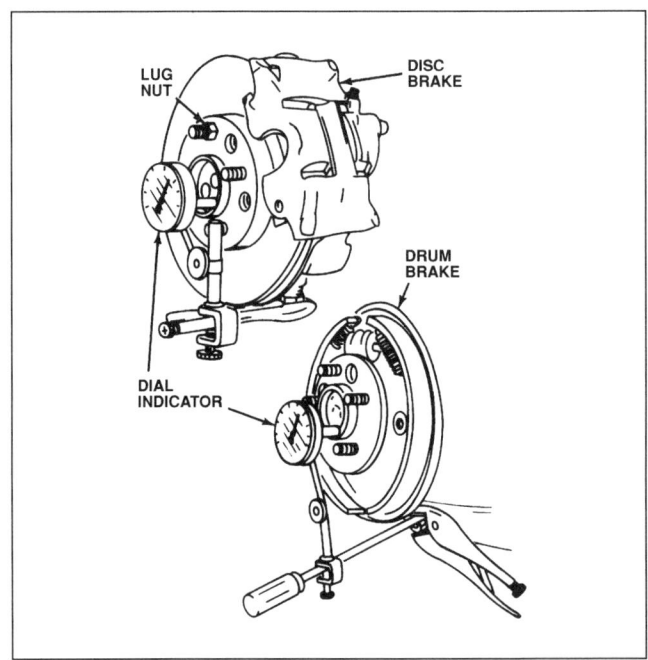

4-1 *Bearing axial play is measured at the edge of the bearing hub on a dial indicator mounted to the suspension.*

Axial Play: Clearance that permits axial motion, or in-and-out-movement, of a part.

---

2. If the bearings to be checked are on a wheel equipped with a disc brake, push the caliper pistons into their bores just far enough that the brake pads do not drag against the rotor.
3. Mount a dial indicator on the suspension and position the plunger against the edge of the bearing hub, figure 4-1.
4. Push the hub inward toward the suspension until it will move no farther. Hold the hub in position and zero the indicator dial.
5. Pull the hub outward away from the suspension until it will move no farther. The dial indicator reading equals the bearing axial play.

Adjustable dual wheel bearings, or **tapered roller bearings**, which are used on the front wheels of most rear-wheel-drive (RWD) vehicles, generally operate with 0.001 to 0.005 inch (0.025 to 0.127 mm) of axial play.

Front-wheel-drive (FWD) vehicles usually have sealed wheel bearings. Typically, sealed ball bearing assemblies are allowed a maximum of 0.002 inch (0.05 mm) of axial play. As with their adjustable counterparts, up to 0.005 inch (0.125 mm) of axial play is acceptable for a sealed tapered roller bearing assembly. Consult the factory shop manual to get the proper specifications. If the axial play of a sealed bearing exceeds the amount allowed by the vehicle manufacturer, replace the bearing.

WHEEL BEARING SERVICE

Whenever you replace a tapered roller or ball bearing that has a separate outer **bearing race**, you must replace the race as well. If you install a new bearing in an old race, it will not fit properly, and premature bearing wear and failure will result. Also, install new grease seals whenever wheel bearings are serviced.

Wheel Bearing Cleaning

When an unsealed wheel bearing is removed, wipe as much old grease as possible off of the bearing, using dry rags or paper towels. Then, inspect the grease on the towels for metal chips or other indications of bearing wear or damage. Clean bearings with fresh petroleum-base solvent. Wash each bearing individually and keep bearings with detachable outer races separated, so they can be assembled in the same races from which they were removed.

Once all of the old grease has been washed out, flush the bearings with a non-petroleum-base brake cleaning fluid; this removes any traces of oil and solvent that can contaminate the new grease and lead to premature bearing failure. Finally, hold the bearings by their cages and dry them with unlubricated, low-pressure compressed air. Direct the air through the bearing so it travels across the rollers from side to side. Never spin a bearing with air while drying it. Spinning it at high speed can cause rapid wear and damage.

Wheel Bearing Inspection

Once the bearings are clean and dry, inspect them for signs of wear and damage, figure 4-2. To inspect a bearing, rotate it carefully in a good light so the complete surface of each ball, roller, and race can be fully checked. If any problems are apparent, or its condition appears questionable, replace the bearing.

Wheel Bearing Lubrication

Always pack a new or used wheel bearing with the type of grease recommended by the vehicle manufacturer. Several different **thickening agents** are used to formulate greases, and most do not mix. Always clean away every trace of old grease before repacking a wheel bearing, and never add new grease to old.

To pack a wheel bearing, work the grease into the cage and races and between the balls or rollers, so that no air spaces remain. The most effective way to do this is to use a bearing packer, which uses air or hydraulic pressure to force new grease through the entire bearing, figure 4-3.

If a bearing packer is unavailable, pack wheel bearings by hand. To hand pack a tapered roller bearing, fill the palm of one hand with grease. Grasp the bearing in your other hand so the large end faces down. Then, draw the bearing across the grease in your palm to force grease into the cage and rollers until it oozes out the opposite side. Repeat this process all around the bearing until it is completely filled with grease. Finish by spreading a medium coating of grease around the outside circumference of the bearing.

Adjustable Dual Wheel Bearing Service

To service a set of adjustable dual wheel bearings, follow these steps:

Tapered Roller Bearing: A bearing assembly that consists of an inner race, tapered cylindrical rollers, a cage to space the rollers apart, and an outer race.

Bearing Race: The portion of a bearing that the rolling elements ride on. Tapered roller bearings usually have a removable outer race that fits into the wheel or brake hub.

Thickening Agent: A component of bearing and chassis grease that retains the oils in the mixture.

BENT CAGE

CAGE DAMAGE CAUSED BY IMPROPER HANDLING OR TOOL USE

GALLING

METAL SMEARS ON ROLLER ENDS CAUSED BY OVERHEATING, OVERLOADING, OR INADEQUATE LUBRICATION

STEP WEAR

NOTCHED WEAR PATTERN ON ROLLER ENDS CAUSED BY ABRASIVES IN THE LUBRICANT

ETCHING AND CORROSION

EATEN AWAY BEARING SURFACE WITH GRAY OR GRAY-BLACK COLOR CAUSED BY MOISTURE CONTAMINATION OF THE LUBRICANT

PITTING AND BRUISING

PITS, DEPRESSIONS, AND GROOVES IN THE BEARING SURFACES CAUSED BY PARTICULATE CONTAMINATION OF THE LUBRICANT

SPALLING

FLAKING AWAY OF THE BEARING SURFACE METAL CAUSED BY FATIGUE

MISALIGNMENT

SKEWED WEAR PATTERN CAUSED BY BENT SPINDLE OR IMPROPER BEARING INSTALLATION

HEAT DISCOLORATION

FAINT YELLOW TO DARK BLUE DISCOL-ORATION FROM OVERHEATING CAUSED BY OVERLOADING OR INADEQUATE LU-BRICATION

BRINELLING

INDENTATIONS IN THE RACES CAUSED BY IMPACT LOADS OR VIBRATION WHEN THE BEARING IS NOT TURNING

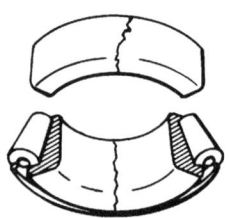

CRACKED RACE

CRACKING OF THE RACE CAUSED BY EXCESSIVE PRESS FIT, IMPROPER IN-STALLATION, OR DAMAGED BEARING SEATS

SMEARING

SMEARED METAL FROM SLIPPAGE CAUSED BY POOR FIT, POOR LUBRICA-TION, OVERLOADING, OVERHEATING, OR HANDLING DAMAGE

FRETTAGE

ETCHING OR CORROSION CAUSED BY SMALL RELATIVE MOVEMENTS BE-TWEEN PARTS WITH NO LUBRICATION

4-2 *Tapered roller bearing inspection guide.*

1. Raise and support the vehicle so the wheels with the bearings to be serviced hang free, then remove the wheels.
2. If the axle is equipped with disc brakes, remove the brake caliper and anchor plate as needed so the rotor can be removed from the axle.
3. Pull the dust cap from the center of the hub to expose the adjusting nut. Remove the cotter pin, retainer, or any other locking devices from the nut. On vehicles that have a split nut with a pinch bolt, loosen the bolt so the adjusting nut can turn freely, figure 4-4.

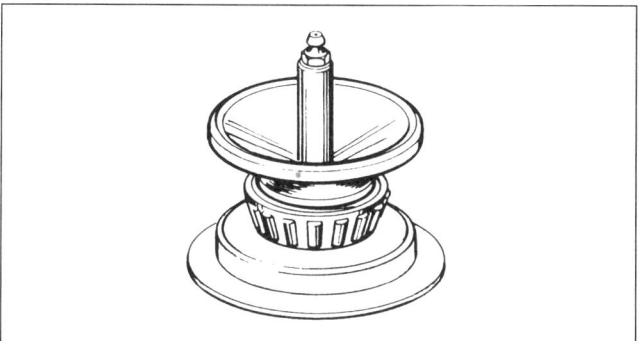

4-3 *A bearing packer uses air or hydraulic pressure to force grease into the bearing.*

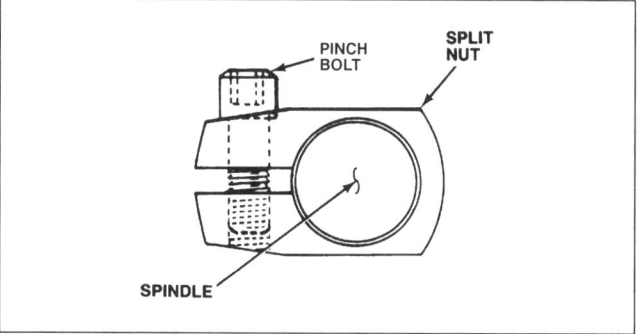

4-4 *On a split nut with a pinch bolt, loosen the pinch bolt so the adjusting nut can turn freely.*

4. Loosen the adjusting nut by backing it off several turns to allow approximately 0.5 inch (13 mm) of play.
5. Pull the drum or rotor outward to free the thrust washer and outer wheel bearing, then push the drum or rotor inward to reseat it on the spindle.
6. Hold the drum or rotor steady to keep it centered, then remove the adjusting nut, thrust washer, and outer wheel bearing from the hub, and set them aside.
7. Pull the drum or rotor straight outward to slide it off the spindle, taking care not to drag the inner wheel bearing across the adjusting nut threads. The brake adjustment may need to be loosened to remove some drums.
8. Use a puller or a pair of pry bars to carefully remove the grease seal and inner bearing.
9. Clean and inspect the bearings and bearing races as described earlier. Also, clean all old grease from the inside of the drum or rotor hub.
10. If installing new bearings, remove the old outer bearing races from the drum or rotor hub. There are two methods of removing bearing races:
    a. With a bearing race puller, figure 4-5.
    b. With a soft-metal (such as brass) drift. Fit the drift through the hub so it firmly contacts the backside of the race. Strike the drift with a hammer while moving it around the race to drive the race from the hub.

11. New races are pressed or driven into the hub with a bearing race driver or a suitably sized socket, figure 4-6. Support the underside of the hub with a block of wood to prevent damage while installing races.
12. Clean and inspect the spindle for rust, scratches, and discoloration. If the spindle is badly scored, cracked, or discolored from overheating, replace it.
13. Lightly coat the spindle with grease.
14. Pack the wheel bearings with grease as described earlier.
15. Place the drum or rotor outer side down on the workbench and lightly coat the inside of the hub with grease to prevent rust.
16. Put a medium coating of grease on the inner bearing race, then place the inner bearing into the race.
17. Use a seal driver to install the grease seal, then apply a light coating of grease on the seal lip.

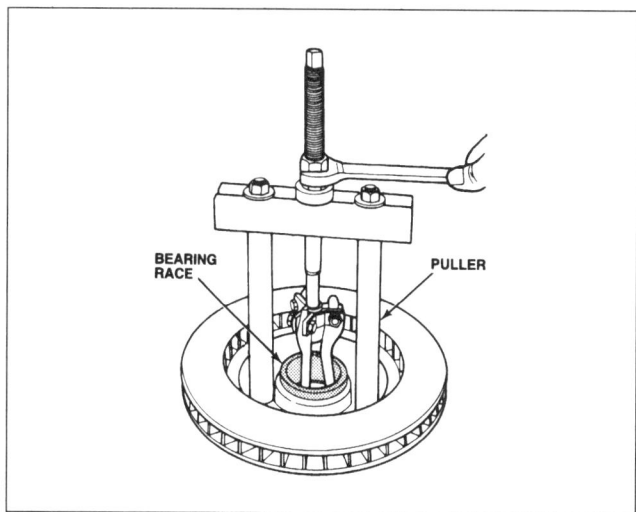

4-5 *Using a bearing race puller to remove the old race from the hub.*

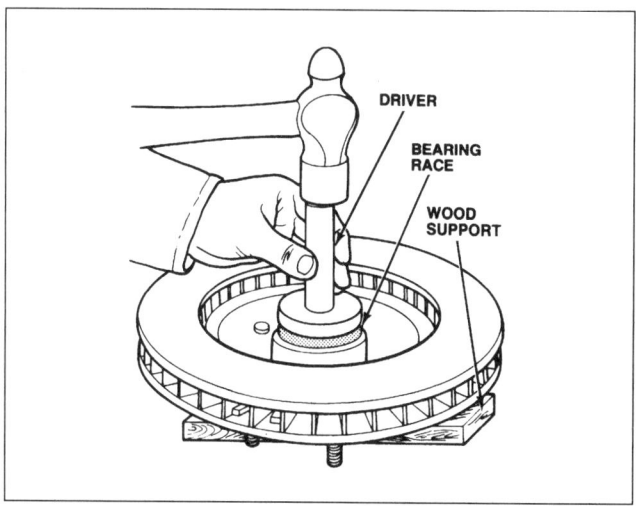

4-6 *Select a suitably sized driver to install bearing races into the hub.*

18. Turn the drum or rotor over and apply a medium coating of grease to the outer bearing race.
19. Fit the drum or rotor squarely over the spindle and slide it straight back into position. Take care to avoid dragging the bearing races across the spindle threads.
20. Hold the drum or rotor in place, fit the outer bearing over the end of the spindle, and slip it into position in the hub. Install the thrust washer over the bearing and thread the adjusting nut onto the spindle finger-tight.
21. Adjust the wheel bearings as described below.
22. If the axle is equipped with disc brakes, install the anchor plate and brake caliper. If the axle has drum brakes and the brake adjustment was loosened, adjust the brakes.
23. Install the wheel and tighten the lug nuts to specified torque following the correct sequence.

Tapered Roller Bearing Adjustment

There are three ways to adjust tapered roller wheel bearings: by hand, with a torque wrench, or using a dial indicator. Once the axial play is properly set, lock the adjusting nut in place, and install the dust cap with a soft-faced hammer. With a castellated adjusting nut, slots on the nut must align with the hole drilled through the spindle, in order to install the cotter pin. If slots are out of alignment after setting axial play, tighten the nut just enough to insert the cotter pin. Do not loosen the nut. When a lock nut is used to secure the adjusting nut, place the lock nut over the adjusting nut so the slots in the lock align with the cotter pin hole in the spindle. With either design, insert a new cotter pin and wrap the tabs around the nut lock or adjusting nut to secure it, figure 4-7. To secure a slotted adjusting nut with a pinch bolt, simply tighten the bolt to the specified torque.

Hand adjustment

To adjust the wheel bearings by hand, rotate the wheel while snugly drawing up the adjusting nut with a wrench

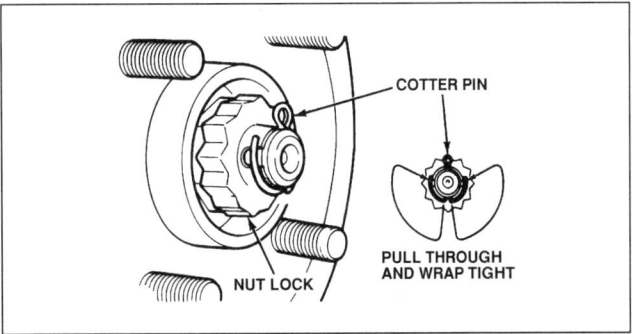

4-7 *Wrap the tabs of the cotter pin around the nut lock or adjusting nut to secure.*

to seat the bearings. Continue to rotate the wheel and back the adjusting nut off 1/4 to 1/2 turn, or until it is just loose. Then, tighten the nut by hand to a snug fit. Check axial play, then lock the adjusting nut in place.

Torque wrench adjustment

To adjust the wheel bearings with a torque wrench, rotate the wheel and draw the adjusting nut up to the initial tightening torque value specified by the vehicle manufacturer, figure 4-8. Typically, tapered roller wheel bearings are initially tightened to about 12 to 25 ft-lb (15 to 35 Nm) of torque. Back off the adjusting nut approximately 1/3 turn, then tighten it to the final tightening torque value specified by the vehicle manufacturer. Final torque typically falls in the 10 to 15 in-lb (1 to 1.5 Nm) range. Check axial play and lock the adjusting nut in place.

Dial indicator adjustment

To adjust the wheel bearings with a dial indicator, tighten the adjusting nut to 12 to 25 ft-lb (15 to 35 Nm) of torque while rotating the wheel. Back off the adjusting nut 1/4 to 1/2 turn or until it is just loose, then tighten the nut by hand to a snug fit. Mount a dial indicator on the wheel and position it so the plunger rests against the end of the spindle, figure 4-9. Push the wheel back

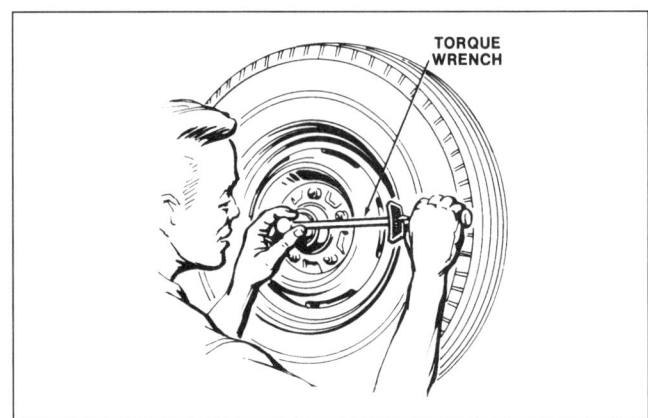

4-8 *Using a torque wrench to adjust tapered roller wheel bearings*

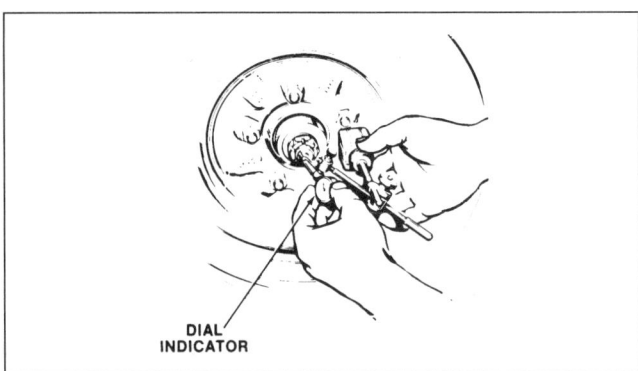

4-9 *Dial indicator setup for adjusting tapered roller wheel bearings*

onto the spindle as far as possible, zero the dial indicator, then pull out on the wheel and read axial play on the dial indicator. Tighten the adjusting nut as needed to obtain the clearance specified by the vehicle manufacturer. Typically, axial play tolerance is in the 0.001 to 0.005 inch (0.025 to 0.127 mm) range. Lock the adjusting nut in place to complete the adjustment.

SEALED WHEEL BEARING SERVICE

The sealed, double-row wheel bearing assemblies used on the front wheels of most FWD vehicles, as well as the driven and non-driven wheels of many late-model vehicles, are serviced by replacing them when their axial play exceeds the recommended specification.

Some sealed rear wheel bearing designs combine the bearings with the wheel hub assembly, which makes replacement a relatively easy procedure, figure 4-10. Simply remove the bearing/hub retaining bolts; then, remove the assembly and replace it.

Replacing a sealed wheel bearing assembly on a driven front axle is more involved. On some designs, the steering knuckle must be removed from the vehicle in order to replace the wheel bearings. A press or special pullers are used to remove the bearings from the steering knuckle, and a press is usually used to install the new bearings, figure 4-11.

Once the new parts are installed and the fasteners are tightened to specified torque, sealed bearings require no adjustment.

PARKING BRAKE INSPECTION

Check parking brake operation during a road test by applying the parking brake while coasting down a gentle hill. The lever or handle should apply smoothly without binding as it slows the vehicle to a stop. If it does not, suspect cable problems. If more than two-thirds of

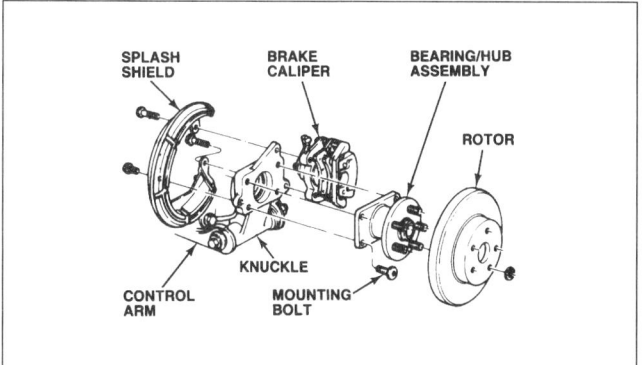

4-10 *The sealed wheel bearings for the driven axle on some RWD vehicles are an integral part of the bearing/hub assembly.*

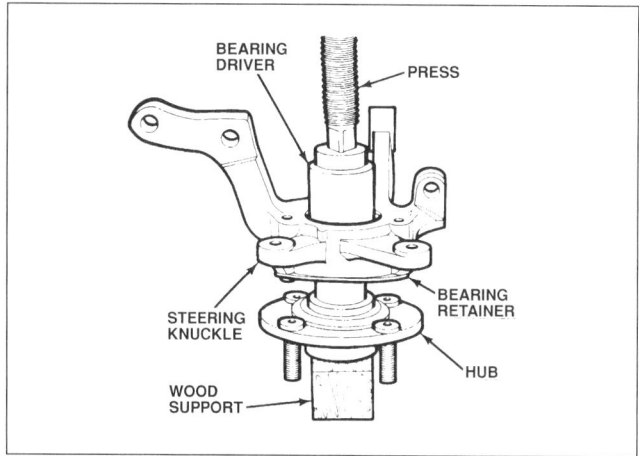

4-11 *The sealed wheel bearings on FWD vehicles are often press fit to the hub and steering knuckle, so the steering knuckle must be removed from the vehicle to service them.*

the parking brake control travel is required to apply the brake, a shoe or cable adjustment is needed. If the parking brake cannot hold the vehicle on a grade, inspect the friction assemblies for glaze on the linings and other problems.

Raise the vehicle on a lift to inspect the parking brake cables. Look for rust, corrosion, and fraying. If the cable is in good condition, lubricate the exposed sections with chassis grease for future protection.

PARKING BRAKE CABLE REPLACEMENT

Parking brake linkages have from one to four cables that can be arranged in a wide variety of configurations. A typical domestic linkage has a control cable that runs from the parking brake control to the equalizer or adjuster, a transfer cable that runs from the equalizer or adjuster to near the rear wheels, and two application cables that run from connectors at the ends of the transfer cable to the parking brake levers at the rear wheels.

When a parking brake cable needs replacement, check the entire run of all the cables to see if any additional parts need replacing along with the cable. Cable mounting hardware is often badly corroded and will break apart when you remove the old cable. If the mounting hardware is in good condition, transfer it to the new cable. After installing the new cable, adjust it as described in the next section, then apply the parking brake hard three or four times to prestretch the cable; readjust the cable if necessary.

PARKING BRAKE ADJUSTMENT

When the parking brake is controlled by a foot pedal or underdash handle, the cable adjuster is generally located under the vehicle at an intermediate lever or equalizer, figure 4-12. If the parking brake is controlled by a floor-mounted lever, the adjustments are usually made inside the passenger compartment where the cables attach to the lever assembly. On most floor-mounted levers, a rubber boot or plastic cover must be lifted up or removed to access the adjustment mechanism, figure 4-13.

Typically, cable adjusters use two jam nuts—an adjusting nut and a lock nut—to set the cable length. However, a few cables designs use a single, self-locking, adjusting nut. If the nuts are seized to the threaded adjuster rod, soak the assembly with penetrating oil before attempting to make the adjustment.

To adjust the cable, hold the adjusting nut in place with an open-end wrench and loosen the lock nut with a second wrench, figure 4-14. Rotate the adjusting nut to draw the end of the cable through the lever or equalizer, to shorten the working length of the cable. Once the adjustment is complete, hold the adjusting nut in place with an open-end wrench and tighten the lock nut against it with a second wrench.

Always use two wrenches to avoid twisting the cables when making an adjustment. Twisting places the cables under additional stress that can lead to premature failure. If necessary, use a pair of locking pliers on an unthreaded section of the threaded rod to hold the cable stationary while tightening the adjusting nut. Some cables have a slot in the end of the threaded rod so a screwdriver can be used to prevent the cable from twisting.

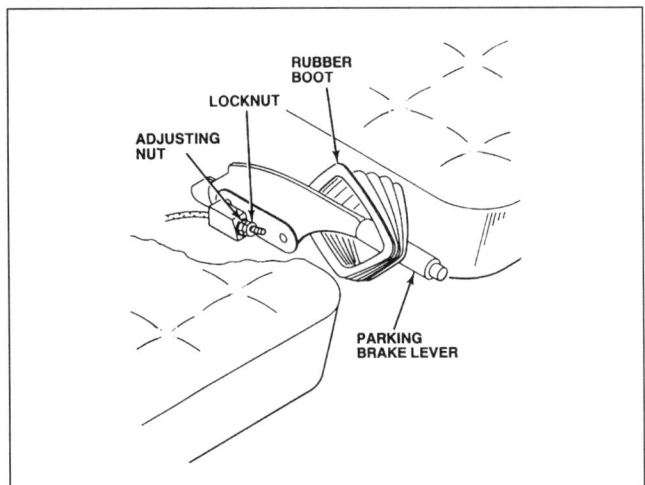

4-13 *Parking brake cables controlled by a floor-mounted lever generally adjust inside the passenger compartment where the cables connect to the lever.*

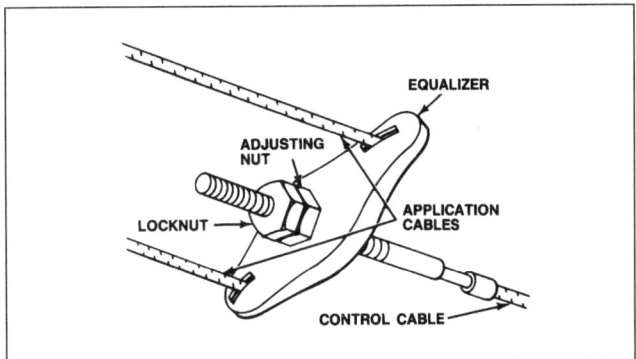

4-12 *Parking brake cables are adjusted at the equalizer, which is located under the vehicle, on most designs that are activated by a foot pedal or underdash handle.*

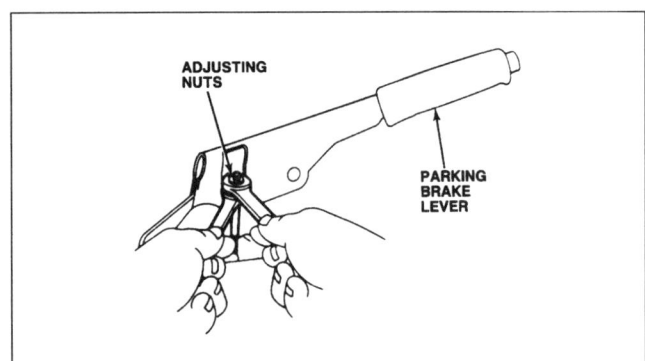

4-14 *Hold the adjusting nut with one wrench and loosen the lock nut with a second wrench, then turn the adjusting nut to obtain the correct cable length.*

1. All of the following indicate that wheel bearings need service *EXCEPT*:
   a. A growling sound that increases with cornering loads
   b. A body vibration that occurs when the brakes are applied
   c. Roughness, grease seal leakage, and excessive axial play
   d. A growling sound that occurs only when the brakes are applied

2. Whether serviceable or sealed, the axial play of a typical tapered roller wheel bearing assembly should be:
   a. 0.001 to 0.005 inch (0.025 to 0.127 mm)
   b. 0.010 to 0.020 inch (0.050 to 0.100 mm)
   c. 0.010 to 0.050 inch (0.50 to 1.0 mm)
   d. Less than 0.002 inch (0.05 mm)

3. Inadequate wheel bearing lubrication can lead to:
   a. Galling
   b. Brinelling
   c. Spalling
   d. Flaking

4. The races of a tapered roller bearing have indentations on their surface as a result of impact loads or vibration when the bearing is not turning. This type of wear is known as:
   a. Spalling
   b. Galling
   c. Brinelling
   d. Frettage

5. Diagnose the condition of sealed, double-row bearings with:
   a. A dial indicator
   b. A seal inspection
   c. A micrometer
   d. A torque wrench

6. Never mix wheel bearing greases because they may have different:
   a. Viscosity improvers
   b. Thickening agents
   c. Operating temperature ranges
   d. Hygroscopic properties

7. Which is not a bearing problem?
   a. Stripping
   b. Spalling
   c. Smearing
   d. Frettage

8. Tapered roller wheel bearings can be adjusted by all of the following *EXCEPT*:
   a. Hand method
   b. Torque wrench method
   c. Tension gauge method
   d. Dial indicator method

9. When parking brakes are properly adjusted, how far through its available travel should the control lever move before the brakes fully apply?
   a. One-quarter
   b. Two-thirds
   c. One-half
   d. Three-quarters

10. The cable adjuster for a foot pedal-activated parking brake is generally found:
    a. Along the rear axle
    b. Inside the car on the control
    c. Under the car at the equalizer
    d. Under the dash panel

Chapter Five

ANTILOCK AND POWER ASSIST SERVICE

Troubleshooting antilock brake problems can sometimes be done without referring to service literature. More often, it requires detailed diagnostic charts found in shop manuals. These charts define the ABS system fault codes and provide step-by-step diagnostic procedures to pinpoint the faulty circuit or component. In addition, ABS testing and service often require a **scan tool** to access the onboard diagnostic programs, as well as for performing service procedures such as system bleeding.

ANTILOCK SERVICE BASICS

Whether ABS equipped or not, most vehicles use essentially the same brake service procedures. This includes techniques for replacing brake pads and shoes, and refinishing rotors and drums. However, some antilock systems may require special brake bleeding procedures.

On vehicles with non-integral, or **add-on ABS**, the basic brake components are usually identical to those used on the same model without the antilock brake option. For example, between the two vehicles, the brake linings, calipers, wheel cylinders, and brake hoses may share the same replacement part numbers. However, the master cylinder may be different on some, but not all, applications. The brake rotors and drums may also be different, depending on whether the wheel speed sensor tone ring is a part of the assembly, or a separate part.

When working on **integral ABS** that uses a pump and **accumulator** rather than a conventional vacuum booster, be sure to vent all pressure from the accumulator before opening any lines or beginning any brake

work, figure 5-1. Pump the brake pedal 25 to 40 times while the ignition is off to relieve pressure. Some ABS applications create hydraulic pressures as high as 2,700 psi (18,660 kPa). Generally, you can monitor the gradual decrease in pressure remaining in the system by sensing the increasing effort required to depress the brake pedal.

Service Precautions

The following general service precautions apply to all vehicles with ABS:

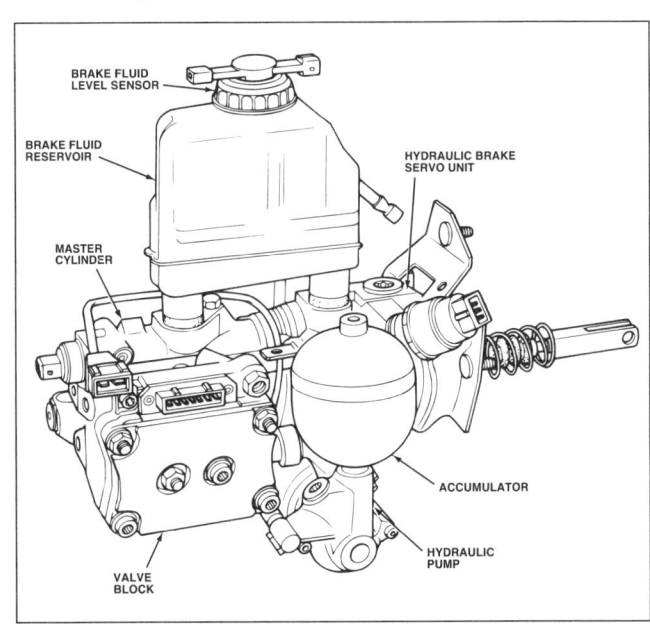

5-1 *On an integral ABS with a hydraulic pump and accumulator, discharge the accumulator by pumping the pedal before opening any lines or beginning any brake work.*

Scan Tool: A diagnostic tool that retrieves diagnostic trouble codes and displays the serial data stream, or input and output signals, of an onboard computer to a technician.

Add-on ABS: An antilock brake system that uses a conventional master cylinder and vacuum brake booster unit in the traditional location on the firewall. The ABS hydraulic modulator is added on elsewhere in the vehicle.

Integral ABS: An antilock brake system that has a single, or integral, hydraulic unit that functions as a master cylinder, power booster, and hydraulic ABS modulator. This hydraulic unit mounts on the firewall in the same location as a conventional master cylinder.

Accumulator: The part of the antilock system that supplies the power assist to apply the brakes. The accumulator is usually a nitrogen-charged reservoir that holds a supply of pressurized brake fluid.

- Always use the proper brake fluid to refill or top off the reservoir. Most systems use DOT 3; none uses DOT 5.
- Never connect or disconnect ABS electrical connectors while the ignition is on. Doing so creates momentary high-voltage spikes that can damage delicate electronic components.
- When testing for open or short circuits, do not ground or apply voltage to any circuit unless instructed to do so by the service manual. Test circuits with a high-impedance digital multimeter (DMM) or a special diagnostic tester only.
- Disconnect the wiring harness from the ABS control module before any type of arc, MIG, or TIG welding is done on the vehicle.
- Do not charge the battery in the vehicle with a high amp fast charger unless the battery cables have been disconnected.
- Heat can damage the ABS control module, so it should be removed before a repainted vehicle is put into a bake oven.
- After replacing an ABS component, check the system thoroughly to make sure it functions correctly.
- Use only top-quality replacement parts in the brake system to assure proper ABS operation.
- With integral ABS, always relieve accumulator pressure before servicing the system. Do this by depressing the brake pedal with a steady force 25 to 40 times while the ignition switch is off. Some imports have a special fitting or plug on the modulator that can be opened to relieve accumulator pressure.

Detecting ABS Problems

In most cases, an ABS malfunction will not affect normal braking. However, there are exceptions, which are explained later. Typically, an ABS problem only affects the ABS portion of the braking system.

Most ABS problems can be detected by monitoring when the ABS warning lamp, BRAKE warning lamp, or both, illuminate. One or both lamps may fail to go out, come on intermittently, or remain on continuously while driving, figure 5-2. Simultaneously, the driver may have noticed a change in the braking characteristics of the vehicle, or a complete loss of ABS function.

Avoid blaming the ABS for conventional brake problems. In general, service grabbing, pulling, dragging, or noisy brakes by following the procedures presented in the preceding chapters. Make sure the service brakes are in good working order before searching for an ABS problem. Remember, most ABS only activates when the **wheel speed sensors** detect a wheel decelerating too quickly, causing an interruption of traction. Integral ABS systems with power assist serve as the exception.

Brake Symptoms

This section describes how the diagnosis of some common brake symptoms may be different with ABS.

- Noise—Most ABS makes noise when operating. Typically, the ABS solenoids in the **modulator** assembly buzz and click. Diagnosis for other brake noises is the same as for conventional brake systems.
- Pulling—Faults with ABS usually do not cause brake pull; pulling is most often the result of a conventional brake system problem. Although unlikely, certain ABS failures can cause the brakes to pull. An ABS isolation valve for a front-wheel circuit that remains energized or stuck closed blocks pressure to the affected wheel. This can cause a pull toward the unaffected side when the brakes are applied. Similarly, an ABS dump valve for a front wheel circuit that remains energized or stuck open prevents the affected brake from being applied and results in a pull to the opposite side. A valve that fails to close is often the result of contamination that interferes with the movement of the valve. Look for debris or corrosion within the hydraulic system when you suspect a sticking ABS valve.
- Pedal vibration—The rapid cycling of brake pressure in the hydraulic circuits during an ABS stop pulsates the brake pedal. The amount of pulsation varies with the type of ABS and vehicle application. Pulsation should only occur during a hard, ABS-assisted stop or when braking on a slick surface. It should not occur during normal braking. If it does, especially when accompanied by a shuddering or jerky stop, check the brake rotors for lateral runout or lack of parallelism. An out-of-round drum, loose wheel bearings, or loose brake parts

---

Wheel Speed Sensor: An electronic sensing device, generally a permanent-magnet generator, that sends information about wheel rotation to the control module of an antilock system. Generally, one sensor is fitted at each of the four wheels.

Modulator: A device that contains high-speed electric solenoid valves that maintain or reduce pressure in the hydraulic brake circuits that feed the calipers or wheel cylinders. The modulator, which is considered heart of an antilock system, prevents wheel lockup.

---

can also cause pedal pulsation during normal braking.

- Grabbing—If the brakes feel jerky or lock up easily during normal braking, check for contaminated linings. Also, examine the drums and rotors for severe scoring. These problems affect ABS operation as well as normal braking. On most antilock systems, low-speed wheel lockup does not indicate a problem. The vehicle must exceed 4 to 6 mph (6 to 10 kph) before the wheel speed sensors provide reliable speed signals and ABS becomes operational.
- Dragging—Although unlikely, ABS with traction control may apply the brake on a drive wheel

continuously if current is constantly applied to the traction control solenoids and pump.

- Gradually sinking brake pedal—This problem is often caused by a worn master cylinder. For systems that have a brake fluid level indicator, check for an illuminated ABS or BRAKE warning lamp. On trucks with Kelsey-Hayes rear wheel antilock brakes (RWAL or RABS), dirt in the ABS control valve can prevent the dump valve from fully seating. This allows brake fluid to leak past the valve causing the pedal to sink.
- Hard Pedal—Increased pedal effort may indicate an ABS problem, but only on integral ABS. The electric pump and accumulator on these systems

SEQUENCE NUMBER	LAMP SEQUENCE	SYMPTOM DESCRIPTION	PERFORM TEST
1		NORMAL LAMP SEQUENCE WITH -EXCESSIVE PEDAL TRAVEL OR SPONGY PEDAL -ANTILOCK BRAKING OPERATION OR VALVE CYCLING DURING NORMAL STOPS ON DRY PAVEMENT -POOR VEHICLE TRACKING DURING ANTILOCK BRAKING	H C D
2		CONTINUOUS "ANTILOCK" LAMP NORMAL "BRAKE" LAMP	A
3		"ANTILOCK" LAMP COMES ON AFTER VEHICLE STARTS MOVING NORMAL BRAKE LAMP	C
4		NO "ANTILOCK" LAMP WHILE CRANKING NORMAL "BRAKE" LAMP	E
5		NO "ANTILOCK" LAMP NORMAL "BRAKE" LAMP	F
6		INTERMITTENT "ANTILOCK" LAMP WHILE DRIVING NORMAL "BRAKE" LAMP	G
7		CONTINUOUS "ANTILOCK" LAMP CONTINUOUS "BRAKE" LAMP	B
8		"ANTILOCK" AND "BRAKE" LAMPS COME ON WHILE BRAKING	B
9		NORMAL "ANTILOCK" LAMP CONTINUOUS "BRAKE" LAMP	B
10		NORMAL OR CONTINUOUS "ANTILOCK" LAMP FLASHING "BRAKE" LAMP	B

5-2 *This lamp sequence chart, taken from a factory service manual, refers you to a specific troubleshooting procedure based on how the warning lamps illuminate.*

provide normal power assist. Problems such as a faulty pump, pump relay, or a pressure loss in the accumulator reduce boost, which increases the pedal effort needed to stop the vehicle. Usually, the ABS warning lamp lights, and the ABS control module deactivates itself. With a hard pedal on vehicles with hydro-boost power brakes, look for a loose power steering pump belt, low fluid level, leaky hoses, or faulty valves in the hydro-boost assembly.

Preliminary Checks

The following ABS checks are made in addition to the preliminary checks previously recommended for conventional brake components. Before looking for ABS faults, check:

- Battery charge—ABS systems require a fully charged battery; open circuit voltage must be over 10 volts, to operate.
- Fuses—Check the ABS control module fuse, main relay fuse, and pump motor fuse. Also, check instrument cluster fuses that could affect the warning lamps.
- Connectors—Check for corroded or loosely installed connections on the following parts: the main relay, pump motor, pump motor relay, pressure switch, main valve, valve block, fluid-level sensor, control module, and wheel speed sensors.
- Grounds—Check for excessive voltage drop across system ground connections, especially those for the control module, pump motor, relay, and hydraulic modulator assembly.

Verifying ABS Operation

When the antilock system is working normally, the ABS warning lamp should illuminate for a few seconds when the ignition is switched on as a bulb check for most vehicles, figure 5-3. During engine cranking, the ABS warning lamp, and possibly the BRAKE warning lamp, will remain on. Once the engine starts, the BRAKE lamp should go out immediately, and the ABS lamp should switch off after a short delay. On vehicles with traction control, the traction control warning lamp generally functions similarly to the ABS lamp. All of the warning lamps should remain off at all other times.

Test the ABS system by driving on a wet or slick surface at 20 to 25 mph (32 to 48 kph) and stopping suddenly. If ABS works, expect to feel feedback in the brake pedal and hear the buzzing and clicking noises mentioned previously. The vehicle should stop in a straight line without skidding or locking up the wheels.

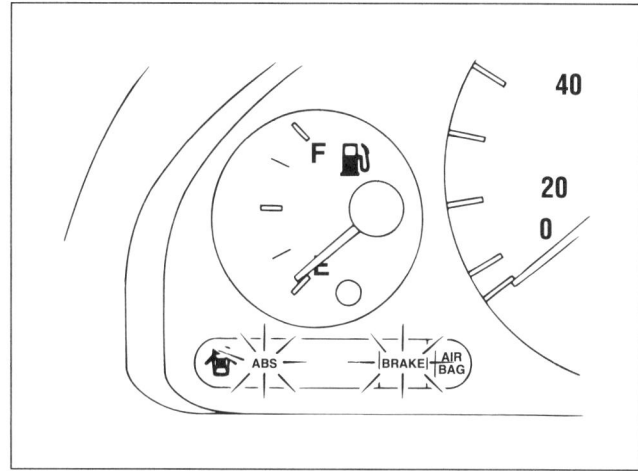

5-3 *The ABS warning lamp, and the BRAKE warning lamp on some models, momentarily illuminates when the ignition is switched on.*

ABS monitoring equipment

A high-impedance, digital multimeter (DMM) is essential for troubleshooting ABS faults. Accurate wiring diagrams and service specifications are needed as well. A scan tool or a dedicated ABS tester is also required for servicing some systems, figure 5-4.

A scan tool can check for diagnostic trouble codes (DTC), as well as display data stream parameters and perform functional tests on some systems, figure 5-5. Functional testing allows you to quickly check the operation of the pump and the modulator solenoids and motors. However, some test functions require bidirectional communication between the scan tool and ABS control module. This allows the scan tool to receive and give commands to the ABS control module. Some manufacturers limit bidirectional communications information and these features are only available with their factory scan tool.

When scan tool testing reveals ABS codes in memory, refer to the appropriate diagnostic chart for troubleshooting. Often, it is necessary to perform pinpoint tests using a DMM and breakout box to isolate the source of the problem.

Diagnosis With the ABS Warning Lamp

Check bulb operation during the timed bulb check period when the ignition switch is initially turned on. If the bulb illuminates and extinguishes after several seconds, and there are no other brake system complaints, then ABS is probably operational. However, the only way to know for sure is to test the system.

If the ABS warning lamp comes on and remains on, the system has detected a fault that requires further

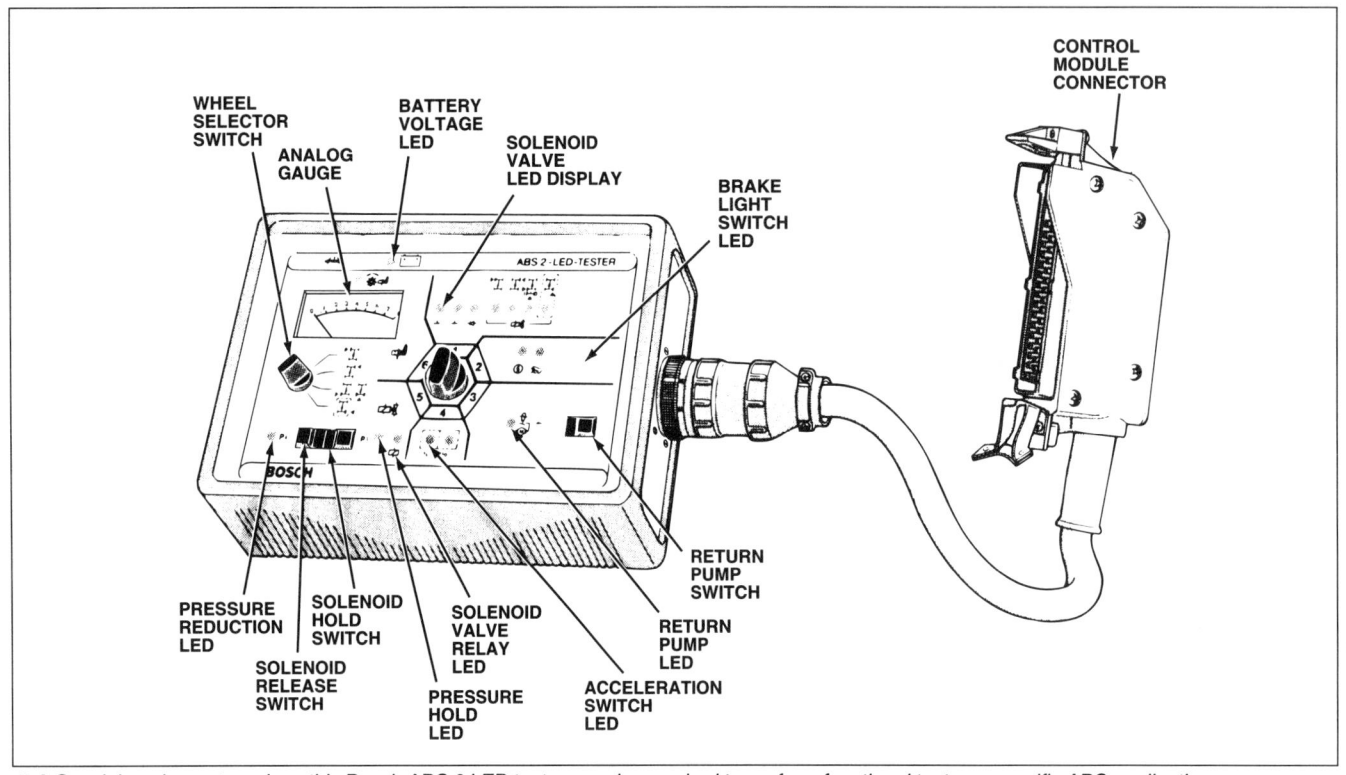

5-4 *Special equipment, such as this Bosch ABS 2 LED tester, may be required to perform functional tests on specific ABS applications.*

diagnosis. A BRAKE warning lamp that remains on or comes on while driving usually indicates a problem with the hydraulic system, not an ABS failure. Be aware: Some systems use the BRAKE warning lamp to alert the driver of an ABS problem when the ABS lamp or circuit is malfunctioning.

If the ABS lamp comes on and remains on, or flashes, the self-diagnostic program of ABS control module has detected a system failure. How the warning lamp reacts can provide clues to the nature of the problem.

An ABS warning lamp that lights when the vehicle first begins to move, generally indicates a problem with

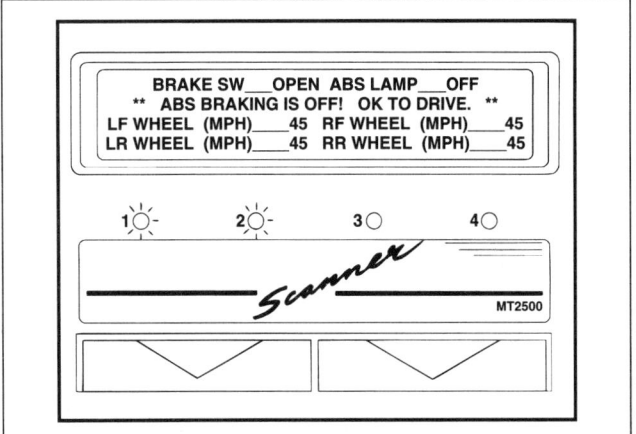

5-5 *A scan tool can retrieve trouble codes, monitor the data stream, and activate functional tests on some ABS applications.*

one wheel speed sensor. Speed sensor failures can also cause ABS to engage during normal stops on dry pavement.

Most systems perform a self-test to check the integrity of the circuits when the ignition is turned on. Once the vehicle gets underway, a second self-test momentarily energizes solenoids, valves, and motors to check for a dynamic response. Should the system detect a fault during either self-test, the warning lamp illuminates and the control module suspends ABS operation.

On vehicles with integral ABS, if both the BRAKE and ABS warning lamps illuminate and power assist is low, suspect an inoperative pump or an accumulator pressure leak. If power assist is normal and both warning lamps are on, check the fluid level and fluid-level sensor. If the level is normal and the sensor is working, take a brake pressure reading.

When ABS is combined with a traction control system, both systems automatically deactivate when the warning lamp comes on as a failsafe procedure. This should not affect normal braking, and does not pose any danger to the safe operation of the vehicle under normal driving conditions. Vehicles with integral ABS also deactivate power-assisted braking.

If the ABS warning light is on and remains on, first check for obvious problems such as a low fluid level. Next, retrieve diagnostic trouble codes. Refer to a diagnostic chart for the specific vehicle being serviced to conduct voltage, resistance, and continuity tests.

ONBOARD DIAGNOSTICS

Most antilock brake systems have comprehensive self-diagnostic capability; early Bosch systems serve as exceptions.

Typically, the ABS control module generates diagnostic trouble codes. Each DTC represents a specific failure, such as a signal loss from a wheel-speed sensor, erroneous voltage feedback from a modulator solenoid, or an intermittent signal from a pump motor relay circuit. On some applications, a low pressure switch detects pressure loss in the accumulator, and a switch circuit failure will also set a DTC. Be aware: The ABS self-diagnostic routine monitors ABS functions only, and will not recognize problems or set codes for the conventional brake system.

On some vehicles, retrieve codes by grounding the ABS control module diagnostic connector, and counting the number of times the ABS warning lamp flashes, figure 5-6. On some others, you access codes by pushing buttons on the control panel of the climate control system in a specified sequence to display results on the vehicle information center or digital speedometer. For some applications, a scan tool must be used to retrieve codes from the ABS control module.

Fault Code Diagnosis

Although a DTC pinpoints the specific faulty circuit, it does not tell you the exact nature of the fault, or exactly where the problem is located. The DTC is simply a guide to point you in the right direction. Additional testing is needed to locate the defective component, connection, or wiring. To identify the problem from your list of possibilities, first refer to the diagnostic procedure found in the shop manual. Each step in a troubleshooting sequence eliminates working components and subsystems, which leads to the source of the failure. Often these charts are quite lengthy and involve a number of continuity, resistance, and voltage checks. Carry out each test in the exact order prescribed. Skipping steps, or taking other shortcuts, can lead to false conclusions. Using the factory procedure, you can efficiently identify the problem.

Multiple diagnostic trouble codes

When there are multiple ABS codes in memory, they display either in numerical sequence or the sequence in which you must repair them. Troubleshooting codes in correct order is important, as one code may be responsible for setting other codes. Follow the steps included in the diagnostic chart in sequence to eliminate all possibilities.

When diagnosing with codes, always attempt to make the codes reappear to be sure the problem still exists. First, record and then clear all the codes present. Inspect the wiring for obvious problems such as loose or corroded connectors. Then test drive the vehicle, or perform a wiggle test, or both. Note if the ABS warning lamp comes back on or if the same codes reappear on your scan tool.

When more than three codes appear, the real problem may be a loose or corroded connection shared by multiple circuits. It is extremely rare to have more than one component fail at the same time. Consult a wiring diagram to determine which parts share common circuits on both the power and ground sides.

False codes

An ABS control module may generate false codes, which are a result of the system software design. For example, a wheel speed sensor code may be set if the

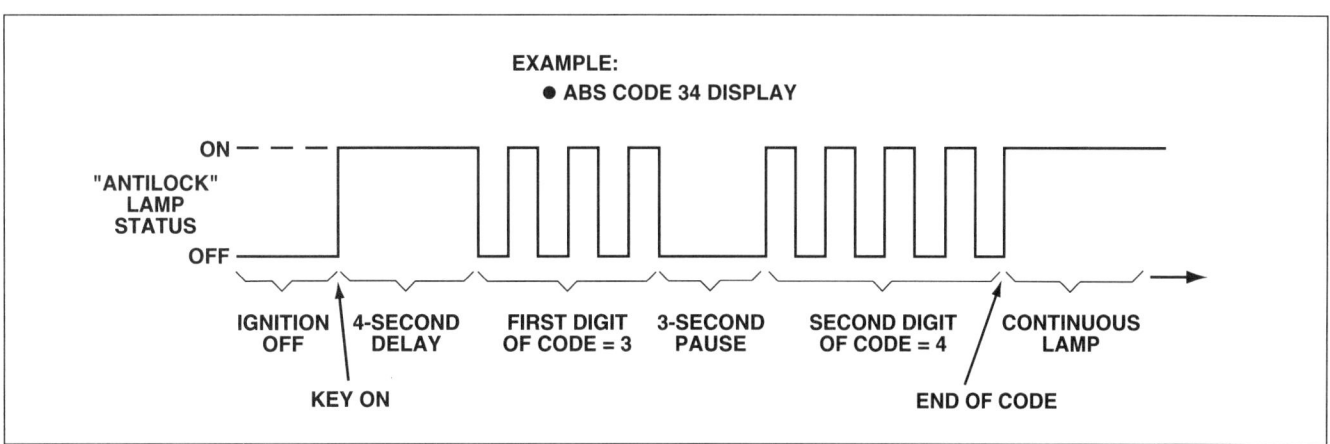

5-6 *Some ABS control modules flash DTC patterns on the warning lamp when placed in diagnostic mode by grounding the test connector.*

rear wheels break traction while accelerating on ice or mud. False wheel speed sensor codes can also be triggered on some vehicles if a wheel rotates or spins while the ignition switch is in the on position. In these cases, the ABS control module sets a code because the speed signal from one wheel disagrees with the signals from the other sensors. Changing a tire or working on the brakes with the ignition switch on can set a false DTC.

Changing tire size can trigger wheel speed sensor codes and cause ABS performance problems as well. If replacement tires or wheels have a larger or smaller diameter than the original equipment, rotational speed, and thus wheel speed sensor voltage, will vary. This can be especially critical if there is a difference in tire diameter between the front and rear axles. Keep in mind, the ABS is calibrated for a specific size of tire for a specific size wheel, and most manufacturers recommend against changing tire sizes.

ANTILOCK INSPECTION AND PERFORMANCE CHECKS

The following are typical inspections and performance checks that apply to systems with integral ABS. On these designs, the master cylinder, pump, accumulator, and modulator are combined in an assembly.

Pump and Accumulator Checks

After relieving accumulator pressure, switch the ignition on to check pump operation. If you do not hear the pump engage, check for voltage available at the pump, and take a voltage drop reading across the ground connection at the pump. Normal voltage available at the pump along with a low voltage drop across the ground connection indicates a defective pump assembly. Replace the pump. A low or zero voltage reading at the pump indicates a problem in the power supply; check the relay and wiring harness. A high voltage drop indicates a problem on the ground side of the circuit.

Check accumulator pressure by connecting a high-pressure gauge between the accumulator and modulator assembly, figure 5-7. Switch the ignition on and the accumulator should quickly pressurize the brake fluid to about 600 to 1200 psi (4137 to 8274 kPa). Then, pressure should slowly climb to peak specified pressure which is sometimes as high as 2700 psi (18,616 kPa). Refer to the appropriate shop manual for exact specifications.

With the gauge connected and the ignition switched on, pump the brake pedal until the pump motor restarts. When the pump stops, switch the ignition off,

wait three minutes, and note the pressure. Wait five more minutes, and note the pressure again. Accumulator pressure leakdown should not exceed 20 psi (138 kPa) in five minutes. If the leakdown is greater, check for leakage in the pump, the master cylinder, or booster assembly.

To locate the pressure leak, switch the ignition on and allow the pump to run for one minute. Then, switch the ignition off and disconnect the return hose from the fluid reservoir. Plug the reservoir outlet and hold the free end of the return hose in a suitable container. Watch the open end of the hose and note what happens during the next five minutes. If there is no fluid flow through the hose, the problem lies in the master cylinder and booster assembly. If fluid flows from the hose, the pump is leaking and the pump and motor assembly should be replaced.

Wheel Speed Sensor Service

Wheel speed sensor circuits are often the cause of ABS problems. These components may suffer from physical damage, buildup of metallic debris on the

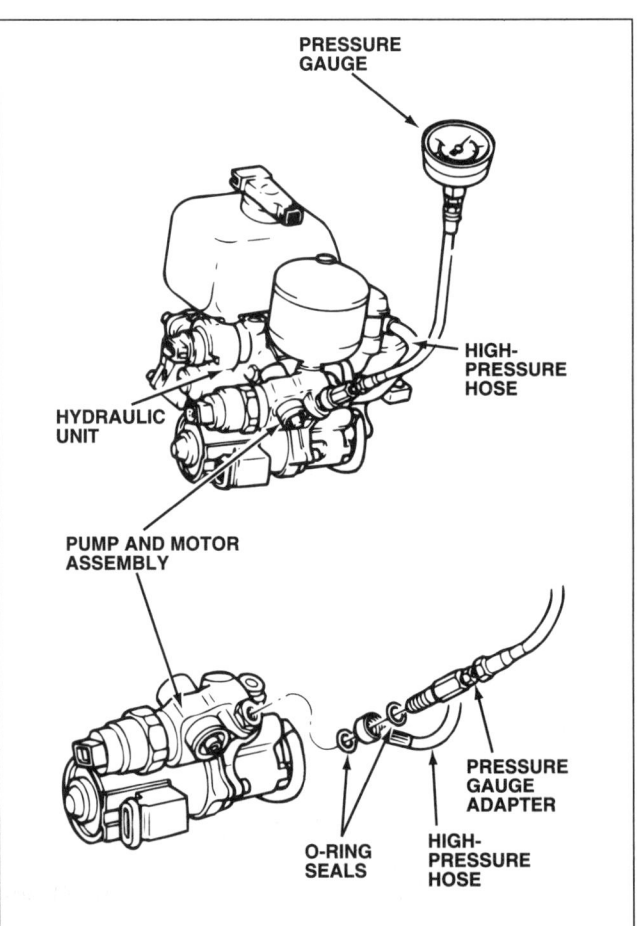

5-7 *Connect a high pressure gauge between the accumulator and modulator assembly to monitor accumulator pressure on an integral system.*

sensor tip, corrosion, poor electrical connections, and damaged wiring.

With the vehicle on a lift, inspect the wheel speed sensors for damage, figure 5-8. The gear wheels should have all their teeth intact, be free of built-up dirt, and the sensor units must be securely mounted. Check the wiring harness connecting the wheel speed sensors to the main harness. Check the wheel bearings. Wear may allow enough wheel wobble to upset the air gap between the speed sensors and the gear wheel, sending inconsistent speed signals to the ABS control module.

Test a wheel speed sensor by measuring its output voltage and circuit continuity with a DMM. Rotate the wheel by hand at a rate of about one revolution per second. Note the voltage reading from the sensor. A functioning wheel speed sensor generally produces an alternating current (AC) voltage that ranges from about 50 to 700 mV. Refer to the shop manual for the exact specifications.

If voltage readings are low, switch the ignition off and check the resistance across the sensor. Expect the value to be between 800 and 1400 ohms; check the shop manual for the exact specifications. If the resistance is out of range, the sensor is shorted or open; replace it. If sensor resistance meets specifications, test the wiring harness for loose or corroded connections, frayed insulation, or other damage.

Locate grounds or shorts in the wheel speed sensor cables by testing for continuity between the wiring connectors. It is best to simply replace defective wiring, rather than repair it by splicing, soldering, or taping.

Wheel speed sensor adjustment

On ABS applications with adjustable wheel speed sensors, always refer to a shop manual for the proper air gap setting. Adjust most sensors by loosening a set screw, then inserting a non-magnetic, plastic or brass, feeler gauge between the tip of the sensor and a high point on the tone ring, figure 5-9. Adjust the position of the sensor so there is a slight drag on the feeler gauge, then tighten the set screw to lock the sensor in place.

When installing new sensors, look for a piece of paper or plastic on the tip end of the unit. This covering must be left in place during installation, as it is the precise thickness to guarantee a correct air gap between the sensor and the tone ring. Adjust the sensor so the tip just touches the tone ring and you can slip the paper or plastic out without ripping it. Tighten the set screw and the air gap is properly set.

Some manufacturers recommend leaving a paper covering in place; the motion of the tone ring removes

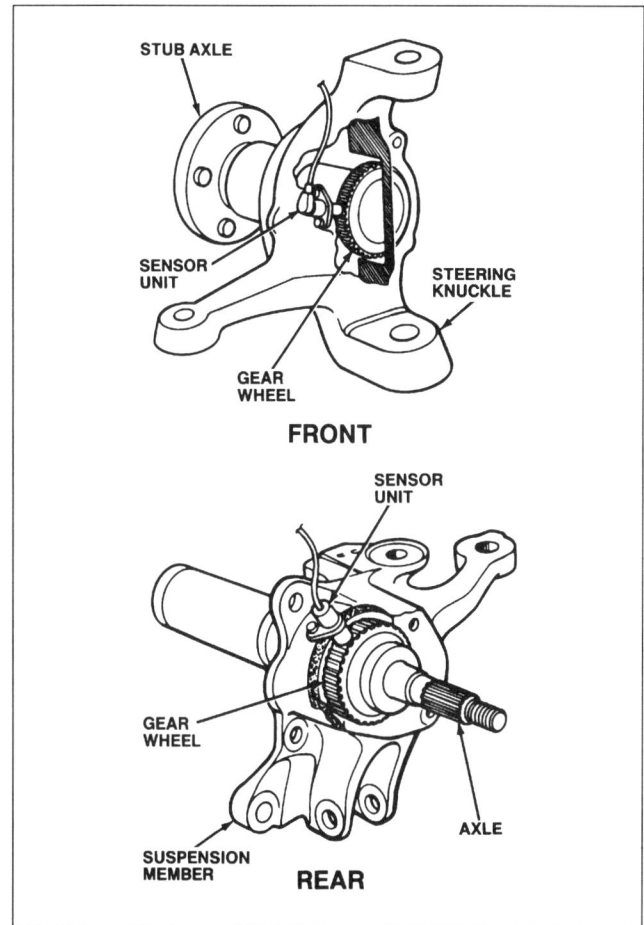

5-8 *Check wheel speed sensors and their tone rings for signs of physical damage and built-up debris that could affect their operation.*

it after the vehicle is driven for several miles. This is the way sensors are generally installed at the factory, and it is not unusual to find traces of the covering still on the sensor when the vehicle is in for service.

When reinstalling a used sensor, be sure that there is no trace of the original paper or plastic covering remaining on the tip. If there is, it will be impossible to properly set the air gap with a feeler gauge. Carefully clean the tip of the sensor to avoid damaging the unit.

Wheel speed sensor replacement

Wheel speed sensors are fragile and must be handled with care. Avoid tapping to force the sensor into place; this can fracture the pickup magnet. On some vehicles, the left rear, left front, right front, and right rear wheel sensors may appear identical, but each is slightly different. These are a set, and individual sensors are not interchangeable. Installing a sensor in the wrong location will affect ABS function.

Coat steel wheel sensor housings with an anticorrosive high temperature lubricant, such as synthetic grease or silicone brake grease, before installation,

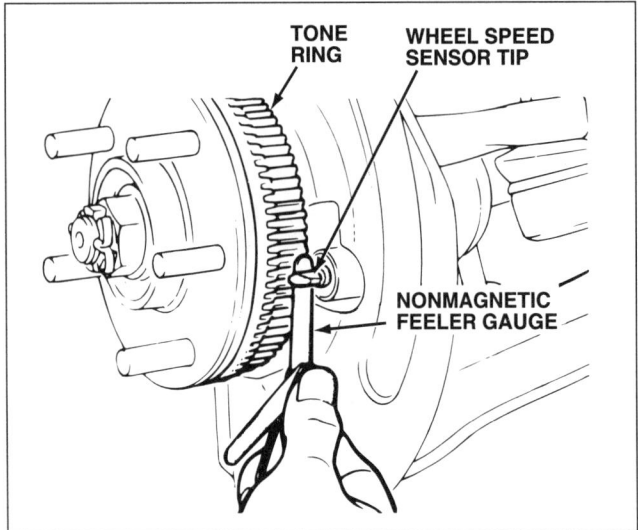

5-9 *Use a non-magnetic feeler gauge to check and adjust the air gap between the wheel speed sensor and the tone ring.*

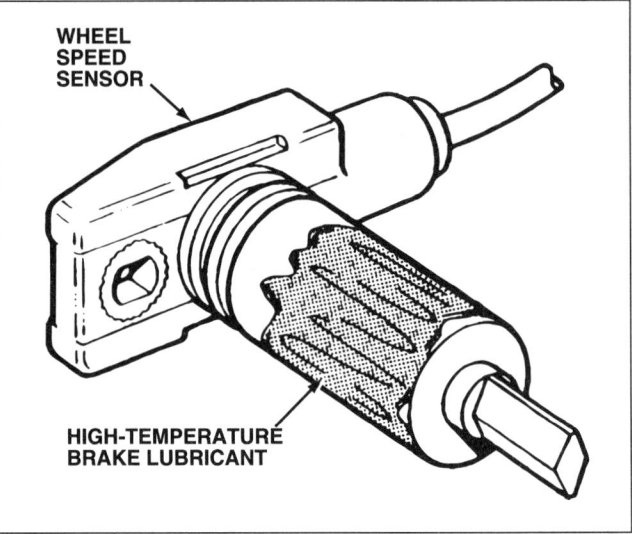

5-10 *Coat steel wheel speed sensors with high-temperature lubricant before installing them.*

figure 5-10. Be sure the lubricant is designed to withstand the high demands of a brake system; do not use ordinary chassis grease. Avoid getting any lubricant on the sensing portion of the assembly; this can result in an erroneous signal to the ABS control module.

Checking the Fluid Level

Check the master cylinder fluid levels in Add-On ABS as you would any vehicle that is not equipped with ABS. Integral ABS is more complex because you check the fluid level with the hydraulic accumulator either charged or discharged, depending on the type of reservoir fitted by the manufacturer.

Charged accumulator

Turn the ignition switch ON, and pump the brake pedal until the hydraulic pump motor begins to run. When the pump stops, the accumulator is fully charged. Check the fluid level against the MAX or FULL mark on the outside wall of the reservoir, figure 5-11.

Discharged accumulator

Turn the ignition switch OFF. Pump the brake pedal as instructed in the Antilock Service Basics section above. Check the fluid levels against the MAX or FULL mark on the outside wall of the reservoir.

ANTILOCK BRAKE BLEEDING

Most Add-On ABS bleeding is identical to that of a conventional brake system, using manual, vacuum, or pressure bleeding. Use the bleeding sequence recommended by the manufacturer. Make sure to include the bleeder valve on the hydraulic modulator or pump if the system is equipped with one. Some modulators have

several bleeder valves and a bleeding sequence of their own.

Although some Integral ABS can be bled like a conventional system, the majority require special bleeding procedures. If the system has a front/rear split hydraulic system, the front wheels can be bled the same as a normal brake system, manually or with a pressure bleeder. Bleed the rear brakes using either a pressure bleeder or the accumulator pressure of the ABS system. To bleed the rear brakes with a pressure bleeder:

1. Charge the pressure bleeder to 35 psi (240 kPa) of pressure.
2. When bleeding the brakes, open one rear bleeder valve for 10 seconds, then close it. Next, open the bleeder valve at the opposite wheel for the same length of time, then close it. Alternate in this way until the fluid from both bleeders is free of air bubbles.
3. When complete, top off the fluid reservoir.

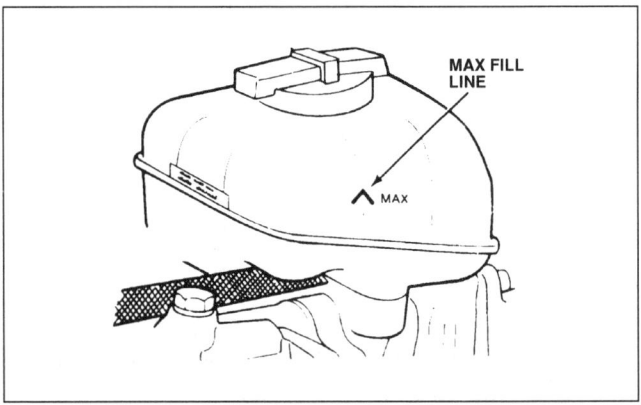

5-11 *The fluid level of integral ABS should be at the MAX or FULL mark on reservoir when the accumulator is fully charged.*

To bleed the rear brakes using accumulator pressure:

1. Turn the ignition switch ON. Press the brake pedal repeatedly until the electro-hydraulic pump motor starts. The motor will stop when the accumulator is charged.
2. Attach one end of a length of plastic tubing over the bleeder valve of the right rear caliper, and submerge the open end of the hose in a container of fresh brake fluid.
3. With the ignition switch on, have an assistant lightly press and hold the brake pedal.
4. Alternately open the rear bleeder valves for 10 seconds at a time until the fluid coming from the bleeders is free of air bubbles. Open the bleeder valve slowly and carefully because the accumulator provides much higher pressure than is available from a pressure bleeder.
5. When bleeding is complete, check and correct the fluid level in the reservoir.

VACUUM POWER ASSIST SERVICE

Most late-model vehicles use a vacuum booster to help apply the brakes. The booster is usually located behind the master cylinder on the firewall. Vacuum boosters require three basic tests:

• Function test
• Vacuum supply test
• Inlet check valve test.

Booster Function Test

Check pedal feel and vacuum booster function while test-driving the vehicle. With the engine off, apply the brake pedal repeatedly with medium pressure until the booster reserve is depleted. At least two brake applications should have a power-assisted feel before the pedal hardens noticeably. If the pedal feels hard immediately, or after only one brake application, it may indicate a vacuum leak or low level of engine vacuum. Inspect the vacuum supply hose to the booster for kinks, cracks, or other damage, figure 5-12.

To test booster function once the reserve is depleted, hold moderate pressure on the brake pedal and start the engine. If the booster is working properly, the pedal will drop slightly.

Booster Vacuum Supply Test

With the ignition OFF, pump the brake pedal to deplete the booster reserve. Disconnect the vacuum supply hose from the booster, and connect a vacuum gauge to the hose using a cone-shaped adapter, figure 5-13. Start the engine and allow it to idle while observing the vacuum gauge. Although the amount of vacuum will vary by application, most will register between 15 and 20 in-Hg (50 and 70 kPa) at idle.

Vacuum Inlet Check Valve Test

To test the vacuum check valve, disconnect the vacuum supply hose from the intake manifold or vacuum pump, and blow into the hose. If air passes through the valve into the booster, the check valve is defective. Replace the check valve.

HYDRO-BOOST POWER ASSIST SERVICE

The Bendix **Hydro-Boost** system is used on vehicles, such as those with Diesel engines, that produce little or no manifold vacuum, figure 5-14. Similar systems are used by several European manufacturers. Hydro-Boost depends on the power steering pump for its boost power, so Hydro-Boost inspection must include checking the power steering pump for leaks, correct fluid level, and drive belt tension. Also, test Hydro-Boost function and accumulator performance.

Hydro-Boost Function Test

With the engine OFF, apply the brake pedal five or more times with medium force to discharge the accumulator. The pedal feel will harden noticeably. Next, apply the brake pedal with medium force, then start the engine. If the booster is working properly, the pedal will drop toward the floor, then push back upward slightly. If the booster passes this test, perform the accumulator test below. However, if there is no change in the

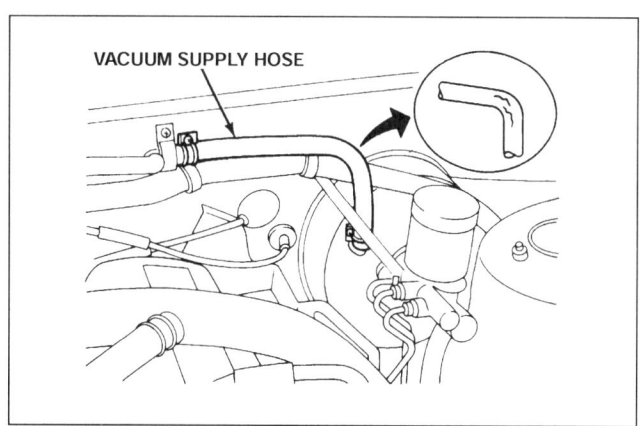

5-12 *Check vacuum booster supply hoses for cracks and damage.*

Hydro-Boost: A mechanical-hydraulic power brake booster. It uses a spool valve to direct hydraulic pressure from the power steering pump to a power piston. The power piston transmits this hydraulic pressure, along with the mechanical force of the driver applying the brake pedal, to the master cylinder.

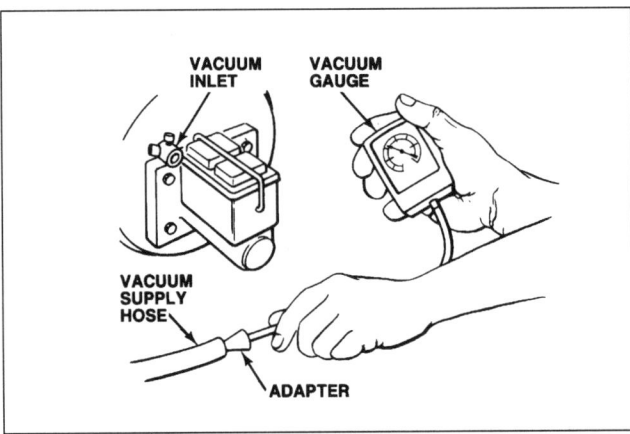

5-13 *Disconnect the vacuum booster supply hose and check for source vacuum with a gauge.*

pedal position or feel, the booster is not working. Check the power steering system to determine whether the problem is in the pump or the booster.

Hydro-Boost Accumulator Test

To test the ability of the system to store a short-term charge, start the engine and allow it to idle. Charge the accumulator by turning the steering wheel slowly one time from lock to lock; do not hold the steering at full lock for more than five seconds. Switch the engine off,

release the steering wheel, and repeatedly apply the brake pedal with medium force. If the accumulator can hold a charge, a Hydro-Boost I unit will provide two or three power assisted applications, while a Hydro-Boost II unit only provides one or two.

To test the ability of the system to store a long-term charge, start the engine and recharge the accumulator as described above. As the accumulator charges on a Hydro-Boost I system, a slight hissing sound should be heard as fluid rushes through the accumulator-charging orifice. Once the accumulator is charged, switch the engine off and do not apply the brake pedal for one hour. At the end of the hour, repeatedly apply the brake pedal with medium force. Once again, a Hydro-Boost I unit should provide two or three power-assisted applications and a Hydro-Boost II unit should provide one or two.

If the Hydro-Boost unit fails these tests, it usually means the accumulator of a Hydro-Boost I unit, or the accumulator/power-piston assembly of a Hydro-Boost II unit, is leaking. In either case, the booster must be rebuilt or replaced. However, if a Hydro-Boost I system fails the test but does not make the hissing sound to indicate the accumulator is charging, the fluid in the system is probably contaminated. Simply flushing the Hydro-Boost system may cure the problem.

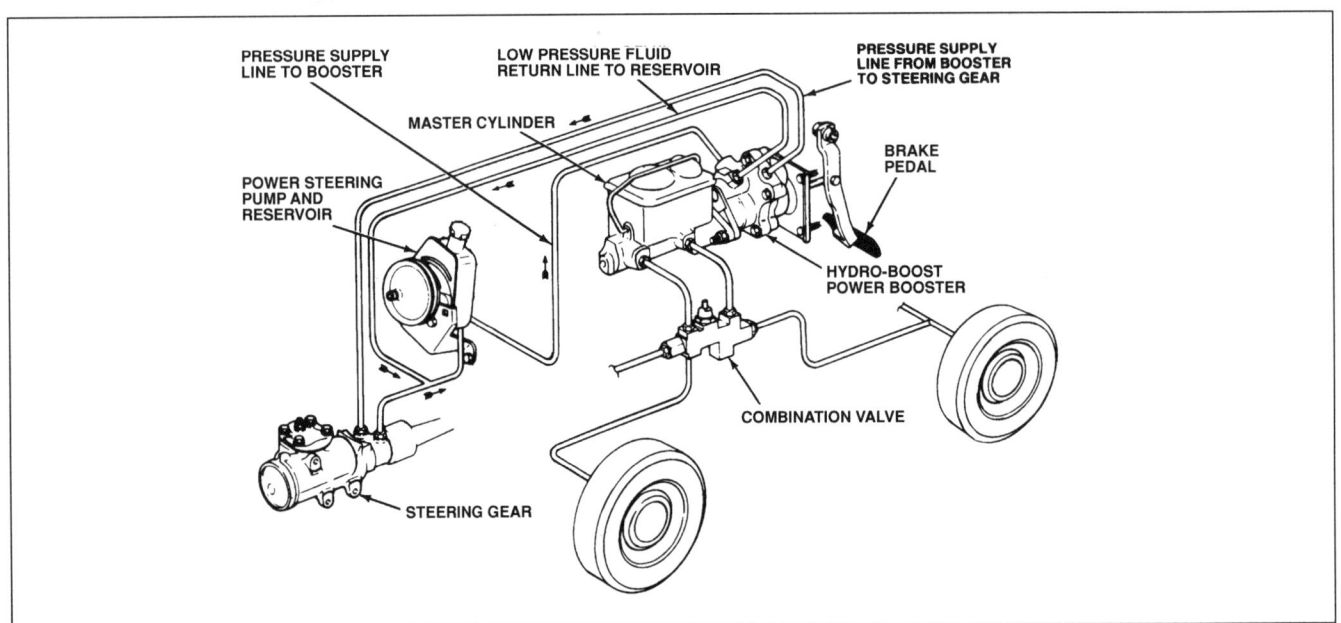

5-14 *The Bendix Hydro-Boost system uses hydraulic pressure from the power steering pump to provide power brake assist.*

1. When first testing a car equipped with ABS, it is important to:
   a. Try to lock the brakes
   b. Disconnect the battery ground
   c. Use the parking brake
   d. Note the status of the ABS warning light

2. To avoid spraying the vehicle with brake fluid when removing a hydraulic part from an integral ABS system, it is important to first:
   a. Bleed the system
   b. Wrap the connection with a shop towel
   c. Relieve system pressure
   d. Siphon fluid from the reservoir

3. An ABS warning lamp that comes on when the vehicle first begins to move generally indicates:
   a. A wheel speed sensor problem
   b. A master cylinder problem
   c. An isolation valve problem
   d. An accumulator problem

4. If both the BRAKE and ABS warning lamps illuminate and power assist is low on an integral ABS system, the probable cause is:
   a. A wheel speed sensor or circuit failure
   b. An inoperative pump or an accumulator pressure leak
   c. Dirt in the ABS control valve preventing the dump valve from seating
   d. A malfunctioning isolation valve or a stuck open dump valve

5. When inspecting wheel speed sensors, check for all of the following *EXCEPT*:
   a. Condition of the tone wheel teeth
   b. Secure sensor mounting
   c. Correct input voltage
   d. Proper air gap between the sensor and tone ring

6. To check the fluid level on an integral antilock system when the accumulator must be charged:
   a. Switch the ignition OFF, pump the brake pedal until the pedal feels hard, then check the fluid level
   b. Check fluid level as you would on a vehicle without ABS
   c. Switch the ignition ON, pump the brake pedal to start the pump motor, then check the fluid level after the pump stops
   d. Switch the ignition ON, pump the brake pedal to start the pump motor, then check the fluid level with the pump running

7. When multiple trouble codes are present in an antilock system, look for:
   a. A weak connection at a common ground
   b. An open circuit
   c. Low-voltage signals
   d. Two or more faulty sensors

8. The peak accumulator pressure developed by an ABS hydraulic pump may be as high as:
   a. 600 psi (4137 kPa)
   b. 1,200 psi (8274 kPa)
   c. 1,800 psi (12,573 kPa)
   d. 2,700 psi (18,616 kPa)

9. To pressure-bleed the rear brakes of an integral antilock system, the bleeder should be charged to:
   a. 25 psi (172 kPa)
   b. 35 psi (241 kPa)
   c. 45 psi (310 kPa)
   d. 55 psi (379 kPa)

10. If a vacuum power booster is in good condition, starting the engine after the booster has been depleted will cause the brake pedal to:
    a. Sink slightly
    b. Pulse rapidly
    c. Rise slightly
    d. Stay the same

11. When you blow into a vacuum check valve:
    a. Air should pass through to the engine freely
    b. Air should not pass through to the engine
    c. Air should pass through with difficulty
    d. Air should only pass through with vacuum applied

12. When performing a Hydro-Boost function test, discharge the booster, start the engine, and the booster is working properly if the pedal:
    a. Pushes back
    b. Pushes back, then sinks toward the floor
    c. Sinks toward the floor, then pushes back
    d. Sinks toward the floor

This sample test will help you review your knowledge of this entire book. The format of the questions is similar to the certification tests given by the National Institute for Automotive Service Excellence. Generally, the questions here are more difficult than the programmed study questions you answered as you read the technical material in this book.

Read these review questions carefully, then read all the possible answers before making your decision. Always select the **best possible answer**. In some cases, you may think all the answers are partially correct, or you may feel that none is exactly right. But in every case, there is a **best** answer; that is the one you should select.

Answers to the questions in this sample test will be found near the end of this book, following the glossary. If you answer at least 20 of these questions correctly, then you can be confident of your knowledge of the subjects covered in this book and in the ASE Certification Test A5 Brakes. If you answer fewer than 20 correctly, you should reread the text and take another look at the illustrations. Also, check the glossary as you review the material.

1. Technician A says that during a master cylinder overhaul, the primary piston is the first piston installed on assembly.
    Technician B says that during a master cylinder overhaul, the secondary piston is the first piston installed on assembly.
    Who is right?
    a. A only
    b. B only
    c. Both A and B
    d. Neither A nor B

2. Technician A says that a feeler gauge and a new piston can be used to check for an oversized master cylinder bore.
    Technician B says that honing a master cylinder bore should leave the bore wall completely smooth.
    Who is right?
    a. A only
    b. B only
    c. Neither A nor B
    d. Both A and B

3. A vehicle has the master cylinder mounted next to the steering column in front of the driver.
    Technician A says that the right front wheel is the first to be bled, followed by the left rear.
    Technician B says that it takes 3 to 5 strokes of the pedal to bleed each cylinder, and the bleeder valve should be closed prior to the down stroke.
    Who is right?
    a. A only
    b. B only
    c. Both A and B
    d. Neither A nor B

4. Technician A says the pressure differential switch may need to be recentered after bleeding the brakes.

Technician B says there are several types of pressure differential switches, and each requires a different method for recentering.
Who is right?
a. A only
b. B only
c. Both A and B
d. Neither A nor B

5. Technician A says the best way to test a proportioning valve is with a feeler gauge.
    Technician B says the best way to test a proportioning valve is with a vacuum gauge.
    Who is right?
    a. A only
    b. B only
    c. Both A and B
    d. Neither A nor B

6. Technician A says a cracked brake hose that is not leaking need not be replaced.
    Technician B says a brake hose with a blister must be replaced.
    Who is right?
    a. A only
    b. B only
    c. Both A and B
    d. Neither A nor B

7. Brake pedal reserve that gradually fades under light pressure indicates:
    a. Leaks in the brake lines
    b. Weak brake pedal return spring tension
    c. Internal leaks
    d. The system is operating properly

8. Which of the following is NOT a method used to force air and contaminated fluid out of a brake system?

a. Manual brake bleeding
b. Pressure tank bleeding
c. Vacuum bleeding
d. Bench bleeding

9. The first component in a brake system bleeding sequence is the:
    a. Master cylinder
    b. Wheel cylinder
    c. Combination valve
    d. Proportioning valve

10. The quick-take-up (QTU) valve on a master cylinder regulates fluid flow:
    a. Between the cylinders and the reservoir
    b. Between the reservoir and the pushrod
    c. Between opposite wheels
    d. Between the two hydraulic circuits of the system

11. Floating drums may be held in place by all of the following ***EXCEPT***:
    a. Speed nuts over the wheel studs
    b. Bolts threaded into the hub
    c. The wheel and lug nuts
    d. A castellated spindle nut

12. To detect an out-of-round brake drum, measure it with a:
    a. Drum micrometer
    b. Drum gauge
    c. Drum dial indicator
    d. Shoe setting caliper

13. After turning a brake drum, there must be sufficient metal remaining so that the drum inside diameter is smaller than the discard diameter by at least:
    a. 0.015 inch (0.40 mm)
    b. 0.030 inch (0.75 mm)
    c. 0.060 inch (1.5 mm)
    d. 0.120 inch (3.0 mm)

14. To bottom multi-piston calipers in their bores, use:
    a. A screwdriver
    b. A C-clamp
    c. Slip-joint pliers
    d. A special tool

15. Most manufacturers specify a caliper piston-to-bore clearance of:
    a. 0.004 to 0.006 inch (mm)
    b. 0.004 to 0.008 inch (mm)
    c. 0.004 to 0.010 inch (mm)
    d. 0.004 to 0.014 inch (mm)

16. A dial indicator can be used to check a brake rotor for:
    a. Radial runout
    b. Parallelism
    c. Taper variation
    d. Lateral runout

17. Technician A says that when turning drums or rotors, both components on the same axle must be machined to the same tolerance.
    Technician B says that the least amount of metal should be removed during machining.
    Who is right?
    a. A only
    b. B only
    c. Both A and B
    d. Neither A nor B

18. Technician A says that a new bearing should never be installed in an old bearing race.
    Technician B says that all wheel bearings are adjustable.
    Who is right?
    a. A only
    b. B only
    c. Both A and B
    d. Neither A nor B

19. Technician A says that loose wheel bearings can cause a vibration when the brakes are applied.
    Technician B says that loose wheel bearings can add to rotor lateral runout.
    Who is right?
    a. A only
    b. B only
    c. Both A and B
    d. Neither A nor B

20. A wheel bearing that needs to be serviced may exhibit all of the following **EXCEPT**:
    a. Roughness
    b. Grease seal leakage
    c. Excessive axial play
    d. Lack of parallelism

21. Technician A says that a basic ABS diagnosis includes checking for a fully charged battery.
    Technician B says that a basic ABS diagnosis includes checking for blown fuses.
    Who is right?
    a. A only
    b. B only
    c. Both A and B
    d. Neither A nor B

22. To bleed the rear brakes of an integral ABS, the pressure bleeder should be charged to:
    a. 15 psi (103 kPa)
    b. 25 psi (172 kPa)
    c. 35 psi (241 kPa)
    d. 45 psi (310 kPa)

23. How many power-assisted applications should the Hydro-Boost II have in reserve if the accumulator is working properly?
    a. 5 or 6
    b. 4 or 5
    c. 3 or 4
    d. 1 or 2

24. The Hydro-Boost pretest inspection includes:
    a. Pressure testing the power steering pump
    b. Checking accumulator reserve
    c. Checking the power steering pump fluid level
    d. Testing Hydro-Boost function

25. Technician A says that ABS wheel speed sensors are not adjustable.
    Technician B says that some ABS wheel speed sensors require adjustment upon installation.
    Who is right?
    a. A only
    b. B only
    c. Both A and B
    d. Neither A nor B

Test Answers

Chapter 1 Review Questions

1. c, 2. d, 3. a, 4. b, 5. b, 6. a, 7. b, 8. b, 9. a, 10. a, 11. d, 12. a, 13. a, 14. b, 15. c

Chapter 2 Review Questions

1. d, 2. d, 3. c, 4. a, 5. c, 6. d, 7. b, 8. d, 9. b, 10. a, 11. c, 12. b, 13. a, 14. d, 15. a

Chapter 3 Review Questions

1. b, 2. c, 3. a, 4. b, 5. a, 6. c, 7. d, 8. b, 9. d, 10. b, 11. a, 12. c, 13. d, 14. c, 15. a

Chapter 4 Review Questions

1. d, 2. a, 3. a, 4. c, 5. a, 6. b, 7. a, 8. c, 9. b, 10. b

Chapter 5 Review Questions

1. d, 2. c, 3. a, 4. b, 5. c, 6. c, 7. a, 8. d, 9. b, 10. a, 11. b, 12. c

Sample Test Questions

1. b, 2. a, 3. d, 4. c, 5. d, 6. b, 7. c, 8. d, 9. a, 10. a, 11. d, 12. a, 13. b, 14. a, 15. c, 16. d, 17. c, 18. a, 19. c, 20. d, 21. c, 22. c, 23. d, 24. c, 25. b

Accumulator: The part of the antilock system that supplies the power assist to apply the brakes. The accumulator is usually a nitrogen-charged reservoir that holds a supply of pressurized brake fluid.

Add-on ABS: An antilock brake system that uses a conventional master cylinder and vacuum brake booster unit in the traditional location on the firewall. The ABS hydraulic modulator is added on elsewhere in the vehicle.

Anodized Finish: An electrolytically applied coating of a protective oxide. It slows wear.

Axial Play: Clearance that permits axial motion, or in-and-out-movement, of a part.

Bearing Race: The portion of a bearing that the rolling elements ride on. Tapered roller bearings usually have a removable outer race that fits into the wheel or brake hub.

Compensating Port: The opening between the master cylinder reservoir and the cylinder bore that allows fluid to enter or exit the hydraulic system to adjust for changes in volume.

Dual-servo Brake: A drum brake that has servo action, or force transmitted by the wheel cylinder, in both the forward and reverse directions. The rear drum brakes on most rear-wheel-drive domestic vehicles are dual servo.

Fixed Caliper: A brake caliper that solidly bolts to the vehicle suspension. The caliper does not move when the brakes are applied.

Fixed Drum: A brake drum that is cast in one piece along with the hub assembly, which contains the wheel bearings. Fixed drums are most often used on non-driven axles, such as at the rear of a front-wheel-drive vehicle.

Fixed Seal: A rubber seal that fits in a groove machined into the caliper bore. The caliper piston slides through the inside of the seal as the brakes apply and release.

Floating Caliper: A two-piece brake caliper consisting of a rigid anchor plate and movable body that compresses the pads as the brakes apply. The caliper body is supported by bushings and O-rings that slide on guide pins and sleeves, which attach to the caliper anchor plate.

Floating Drum: A brake drum that installs on a separate axle flange or hub assembly. Floating drums, which are commonly found on the drive axle of rear-wheel-drive vehicles, can also be used on a non-driven axle.

Floating Rotor: A floating rotor is a separate part that installs onto an axle flange or hub assembly. Floating rotors are generally held in place by the wheel lug nuts or bolts. The rotor does not contain the wheel bearings, which are part of the hub.

Glaze: A condition that occurs on the surface of brake lining as a result of overheating. A glazed lining takes on a smooth, shiny appearance.

Hydro-boost: A mechanical-hydraulic power brake booster. It uses a spool valve to direct hydraulic pressure from the power steering pump to a power piston. The power piston transmits this hydraulic pressure, along with the mechanical force of the driver applying the brake pedal, to the master cylinder.

Hygroscopic: Water-absorbing. Polyglycol brake fluids are hygroscopic.

Integral ABS: An antilock brake system that has a single, or integral, hydraulic unit that functions as a master cylinder, power booster, and hydraulic ABS modulator. This hydraulic unit mounts on the firewall in the same location as a conventional master cylinder.

Leading-trailing Brake: A non-servo brake with one leading shoe and one trailing shoe.

Modulator: A device that contains high-speed electric solenoid valves that maintain or reduce pressure in the hydraulic brake circuits that feed the calipers or wheel cylinders. The modulator, which is considered the heart of an antilock system, prevents wheel lockup.

Primary Shoe: The shoe in a servo brake that transfers a portion of its stopping power to the secondary shoe.

Quick-take-up (QTU) Master Cylinder: A type of master cylinder that applies a large volume of fluid on the initial brake application to take up the clearance designed into low-drag brake calipers.

Quick-take-up (QTU) Valve: The part of the QTU master cylinder that controls the fluid flow between the reservoir and the cylinder bore.

Scan Tool: A diagnostic tool that retrieves diagnostic trouble codes and displays the serial data stream, or input and output signals, of an onboard computer to a technician.

Secondary Shoe: The shoe in a servo brake that receives extra application force from the primary shoe. The lining of a secondary shoe is larger than that of the primary shoe because it does most of the braking.

Shoe Web: The portion of the brake shoe below the lining table that receives the application force from the wheel cylinder.

Sliding Caliper: A two-piece brake caliper consisting of a body and anchor plate. The anchor plate rigidly attaches to the vehicle suspension, and the body slides on machined ways to bring the pads into contact with the rotor.

Speed Nut: A spring-steel clip that threads onto a stud or bolt to position a part. Speed nuts are used to hold floating drums and rotors in place during vehicle assembly.

Stroking Seal: A rubber seal that fits into a groove on the caliper piston and strokes, or moves, with the piston as the brakes apply and release.

Tapered Roller Bearing: A bearing assembly that consists of an inner race, tapered cylindrical rollers, a cage to space the rollers apart, and an outer race.

Thickening Agent: A component of bearing and chassis grease that retains the oils in the mixture.

Turning: A machining process that uses a brake lathe to remove metal from drums and rotors to refinish their friction surfaces.

Ways: Polished, machined surfaces that permit movement between two metal parts. Ways machined into the anchor plate and caliper body that provide a sliding surface for the caliper.

Wheel Speed Sensor: An electronic sensing device, generally a permanent-magnet generator, that sends information about wheel rotation to the control module of an antilock system. Generally, one sensor is fitted at each of the four wheels.